Power Plant Siting

Power Plant Siting

John V. Winter
State of Ohio

David A. Conner
University of Louisville

Van Nostrand Reinhold Environmental Engineering Series

VNR VAN NOSTRAND REINHOLD COMPANY
NEW YORK CINCINNATI ATLANTA DALLAS SAN FRANCISCO
LONDON TORONTO MELBOURNE

Van Nostrand Reinhold Company Regional Offices:
New York Cincinnati Atlanta Dallas San Francisco

Van Nostrand Reinhold Company International Offices:
London Toronto Melbourne

Library of Congress Catalog Card Number: 78-8169
ISBN: 0–442–29554–5

Manufactured in the United States of America

Published by Van Nostrand Reinhold Company
135 West 50th Street, New York, N.Y. 10020

Published simultaneously in Canada by Van Nostrand Reinhold Ltd.

15 14 13 12 11 10 9 8 7 6 5 4 3 2 1

Library of Congress Cataloging in Publication Data

Winter, John V.
Power plant siting.

(Van Nostrand Reinhold environmental engineering series)
Bibliography: p.
Includes index.
1. Electric power plants–United States–Location.
I. Conner, David A., joint author. II. Title.
TK1193.U5W55 333.7 78-8169
ISBN 0442-29554-5

To our Wives

Tina and Jerry

Van Nostrand Reinhold Environmental Engineering Series

THE VAN NOSTRAND REINHOLD ENVIRONMENTAL ENGINEERING SERIES is dedicated to the presentation of current and vital information relative to the engineering aspects of controlling man's physical environment. Systems and subsystems available to exercise control of both the indoor and outdoor environment continue to become more sophisticated and to involve a number of engineering disciplines. The aim of the series is to provide books which, though often concerned with the life cycle–design, installation, and operation and maintenance–of a specific system or subsystem, are complementary when viewed in their relationship to the total environment.

The Van Nostrand Reinhold Environmental Engineering Series includes books concerned with the engineering of mechanical systems designed (1) to control the environment within structures, including those in which manufacturing processes are carried out, and (2) to control the exterior environment through control of waste products expelled by inhabitants of structures and from manufacturing processes. The series includes books on heating, air conditioning and ventilation, control of air and water pollution, control of the acoustic environment, sanitary engineering and waste disposal, illumination, and piping systems for transporting media of all kinds.

Van Nostrand Reinhold Environmental Engineering Series

ADVANCED WASTEWATER TREATMENT, by Russell L. Culp and Gordon L. Culp

ARCHITECTURAL INTERIOR SYSTEMS—Lighting, Air Conditioning, Acoustics, John E. Flynn and Arthur W. Segil

SOLID WASTE MANAGEMENT, by D. Joseph Hagerty, Joseph L. Pavoni and John E. Heer, Jr.

THERMAL INSULATION, by John F. Malloy

AIR POLLUTION AND INDUSTRY, edited by Richard D. Ross

INDUSTRIAL WASTE DISPOSAL, edited by Richard D. Ross

MICROBIAL CONTAMINATION CONTROL FACILITIES, by Robert S. Rurkle and G. Briggs Phillips

SOUND, NOISE, AND VIBRATION CONTROL (Second Edition), by Lyle F. Yerges

NEW CONCEPTS IN WATER PURIFICATION, by Gordon L. Culp and Russell L. Culp

HANDBOOK OF SOLID WASTE DISPOSAL: MATERIALS AND ENERGY RECOVERY, by Joseph L. Pavoni, John E. Heer, Jr., and D. Joseph Hagerty

ENVIRONMENTAL ASSESSMENTS AND STATEMENTS, by John E. Heer, Jr. and D. Joseph Hagerty

ENVIRONMENTAL IMPACT ANALYSIS: A New Dimension in Decision Making, by R. K. Jain, L. V. Urban and G. S. Stacey

CONTROL SYSTEMS FOR HEATING, VENTILATING, AND AIR CONDITIONING (Second Edition), by Roger W. Haines

WATER QUALITY MANAGEMENT PLANNING, edited by Joseph L. Pavoni

HANDBOOK OF ADVANCED WASTEWATER TREATMENT (Second Edition), by Russell L. Culp, George Mack Wesner and Gordon L. Culp

HANDBOOK OF NOISE ASSESSMENT, edited by Daryl N. May

NOISE CONTROL: HANDBOOK OF PRINCIPLES AND PRACTICES, edited by David M. Lipscomb and Arthur C. Taylor

AIR POLLUTION CONTROL TECHNOLOGY, by Robert M. Bethea

POWER PLANT SITING, by John V. Winter and David A Conner

DISINFECTION OF WASTEWATER AND WATER FOR REUSE, by George C. White.

Preface

The term "power plant siting" musters a wide range of responses in the minds of the general public. To some people, the term produces the concept of a highly confused legal proceeding. To some people, the term signals the possibility of new employment and economic expansion. To some people, the term evokes the vision of a mammoth construction project, managed by monopoly interests, that is destroying large segments of the environment. To other people, the term has no meaning.

The range of responses found in the many sectors of society that are involved in the power plant siting process are just as diverse. Many participants, including legislators, regulators, engineers, and environmentalists, are unaware of (1) the detail required, (2) the regulation imposed, or (3) the intricacy involved in the siting of a new power plant. In a like manner, the public is often unaware of its many opportunities to take part in the siting process.

This book attempts to break down and describe the complexities of the siting process in a manner that will be understandable to the uninformed. An attempt is made to view the process from the perspectives of (1) the utility, (2) the governmental regulating agencies, and (3) the concerned public. It is hoped that this book will assist the reader in grasping the immense detail required to site a power plant.

Many people contributed significantly during the evolution of this work. The authors wish to thank Dr. John Williams, Administrator of the Wisconsin Public Service Commission, who provided valuable information and encouragement, David Ross of the Ohio Department of Natural Resources, who made available his many years of experience on biological processes, Ron Yerian of the Ohio Department of Health, who provided fresh perspectives at several key points, Dr. Harold Kohn of the Ohio Environmental Protection Agency, who shared his expertise on problems of the electric industry and who graciously provided editorial advice and constructive criticism, Eugene Plock of the University of Louisville, who shared his expertise on emerging energy-related technologies, Frank Croft of the University of Louisville, who prepared the graphics, Mrs. Roxie Zellers, who assisted in typing draft manuscripts, and Mrs. Donna Greenwell, who assisted in typing the final manuscript.

Three other people deserve special mention. Dr. Joseph Hagerty of the University of Louisville should be recognized as the individual who induced the authors to write this book. Our wives, Tina and Jerry, should be recognized for providing the love and encouragement required to make this book a reality. Jerry is especially thanked for her patience in correcting the authors' many grammatical and spelling errors and for her patience in proofing the manuscript.

John V. Winter
David A. Conner

Contents

LIST OF FIGURES

LIST OF TABLES

Power Plant Siting

1

An energy overview

ENERGY

Since the 1973 Arab oil embargo, the citizens of the United States have become increasingly aware of their energy situation.[1] Initial emphasis came with a presidential proclamation of a national goal of energy independence. Public awareness has since increased through exposure to the numerous major studies and forecasts of energy supply and demand that have been promulgated and published by governmentally and industrially sponsored organizations.[2, 3, 4, 5] This awareness heightened with ever-increasing energy costs.

In this post-1973 era serious questions have been raised concerning the status of the development of alternate energy sources and the status of energy usage patterns within the country. Action has resulted on several fronts. Legislation has been passed to promote the development of new resources. Conservation of present resources has been stressed. And, industry has increased research and development expenditures in energy related areas. In the interim, however, the federal government has softened its thrust for energy independence.[6, 7, 8, 9]

Unfortunately, the average citizen's contact with the energy industry has been limited to his or her exposure to media coverage and his or her experience with the increased cost of electricity, gasoline, and natural gas. Because of this limited exposure, the average citizen is often unable to understand the power plant siting process. In an attempt to insure that the reader of this book has an adequate understanding of the energy setting for the power plant siting process, this chapter will be devoted to an overview of the energy story.

Dependency on Oil

Most of the data gathered in recent energy studies have shown energy equivalents based on millions of barrels of oil. Carl Solberg, in his book *Oil Power,* has given insight on this phenomenon.

> Oil has resettled our population, elected our Presidents, swayed our foreign policy, and legislated our morality. Oil has not only fueled our wars, but is equally the sinew of our might in peacetime: it takes oil to be a super power.[10]

Because of the importance of oil, much of the data presented in this chapter will be represented in oil equivalents. A second common unit of comparison that will be used is the Btu (British Thermal Unit), a measure of energy. (Appendix I contains several energy conversions and fundamental points of energy information.)

In the years before our present dilemma, the United States' energy consumption was directly associated with the growth of the Gross National Product (GNP) as shown in Figures 1-1 and 1-2. As national output increased, a corresponding increase occurred in the rate of energy usage. However, the internal level of production did not increase at the same rate. The difference was made up primarily by imported oil. (Figures 1-3 and 1-4 display domestic production and consumption statistics in a graphical format.)

In 1973 the United States began to ship emergency arms to Israel. As a result, the major petroleum import sources (the countries which

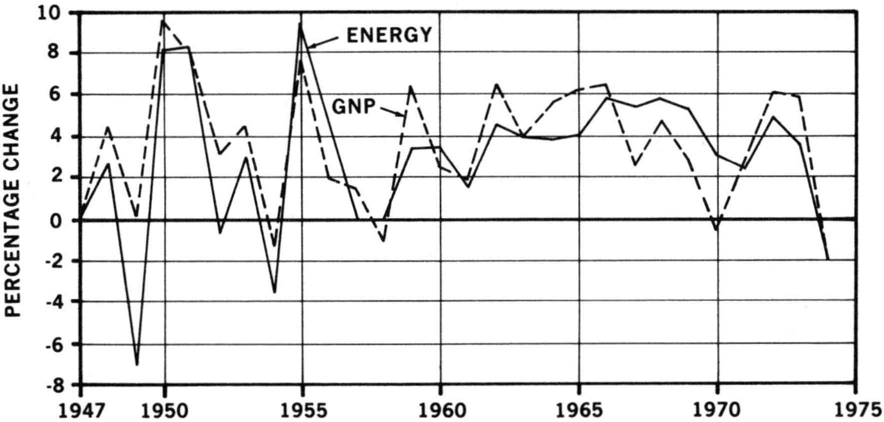

Figure 1-1 Changes in energy and GNP.[3]

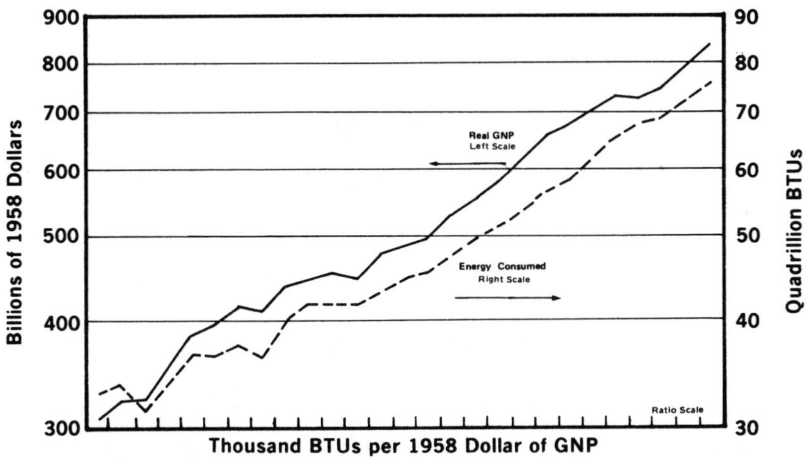

Figure 1-2 Energy and GNP for the United States.[3]

have since become known as the Organization of Petroleum Exporting Countries or OPEC) restricted the export of their oil to the United States. This political action resulted in drastic increases in domestic prices.

Due to increasing oil prices and concurrent efforts to reduce energy consumption, the United States is now consuming energy at a rate

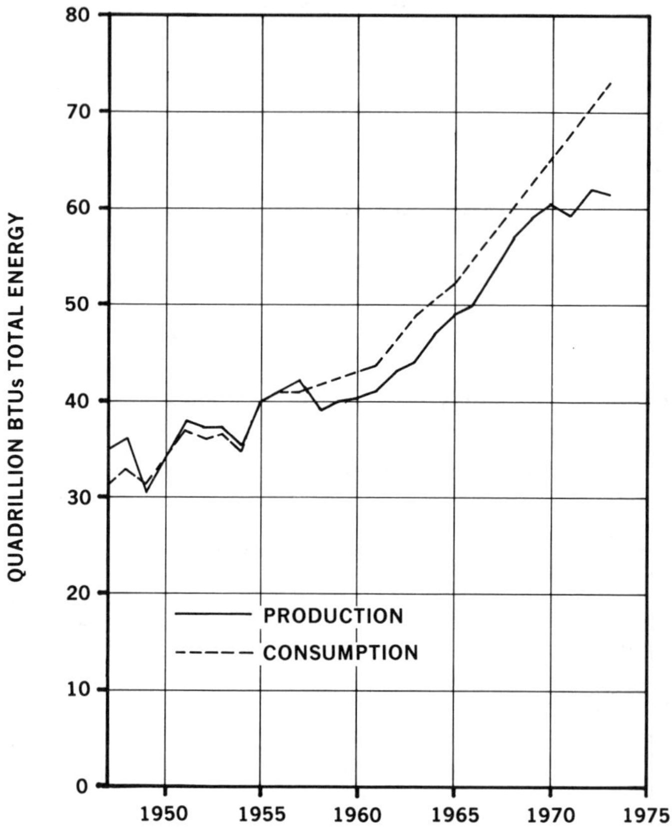

Figure 1-3 United States energy production and consumption referenced to Btu.[6]

that is the equivalent to 40 million barrels of oil per day. Of this consumption, energy equivalent to 17 million barrels of oil per day is being consumed in the form of crude oil. The remaining energy is consumed as natural gas, coal, nuclear energy, and synthetic fuels. The total energy imports are in the five to six million barrel per day range. Where is this energy going?

Energy Statistics — Transition from the Past

Low cost energy prior to 1973 had a direct influence on the "American way of life." Figure 1-5 clearly illustrates this fact. Until 1973

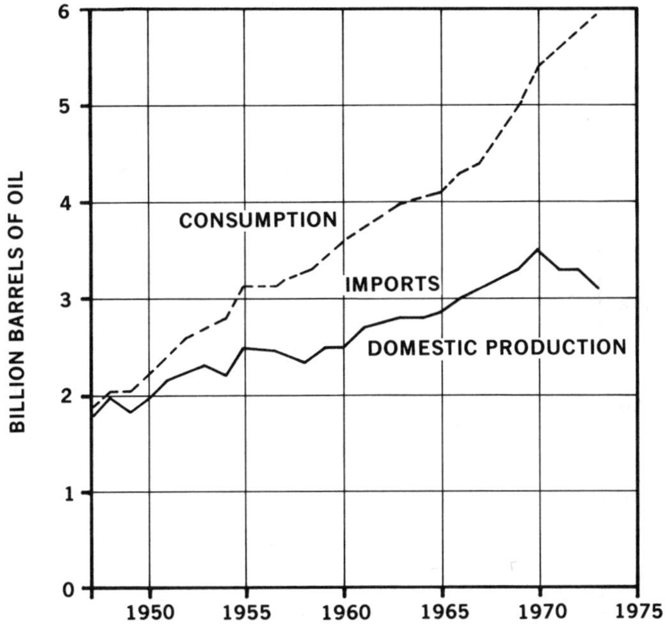

Figure 1-4 United States energy production and consumption referenced to barrels of oil.[6]

the average energy growth rate approached 4% per year.[11] The major increase in energy consumption came from the residential/commercial, industrial, agricultural, and transportation sectors. (Figure 1-6 presents the percentage breakdown of energy consumed in the past few years by usage areas.)

During the past decade, residential energy consumption has accounted for 20% of the total energy consumed. This energy has been utilized to light, heat, and cool homes and to operate major appliances. As energy costs have risen, conservation measures have been taken to reduce heating and cooling energy consumption. In addition, the efficiencies of home appliances have been questioned with emphasis being placed upon reducing their energy consumption.

Energy consumption by the commercial sector is also changing. During the past decade a trend toward "sealed offices" developed, resulting in a need for higher lighting levels and twenty-four hour mechanical ventilation. Extensive use of computers, elevators, and

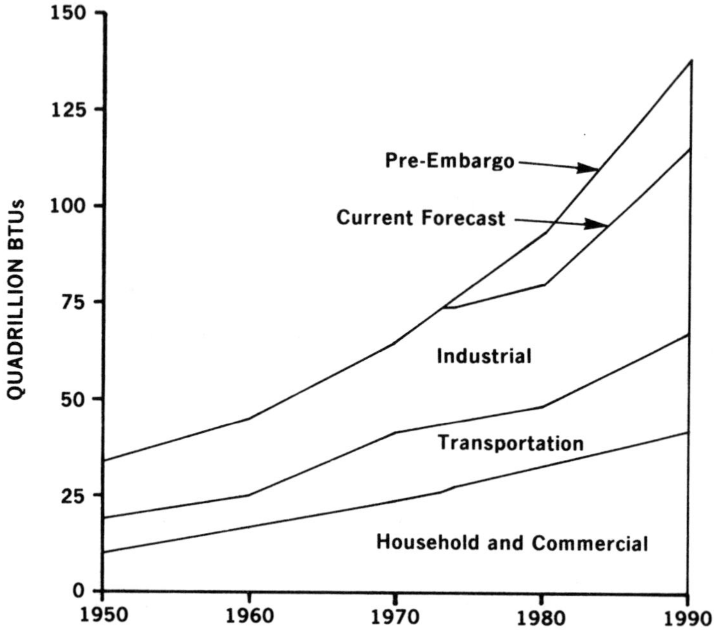

Figure 1-5 United States energy consumption projections based upon past history.[7]

escalators accompanied this "modernization" trend. Now, consideration is being given to facility designs that are less energy intensive.

Throughout the industrial sector, energy has been viewed as a cheap raw material. Energy recovery and conservation have become lost arts. Steam has been vented rather than recycled. Heavy energy consuming machinery has been allowed to idle for long periods rather than being turned off. Inefficient designs have proliferated. Most industry is now conducting internal audits to determine where energy savings can be effected and to establish a dollar value for these savings. Programs are being initiated to increase employee awareness in reducing energy consumption.

Agriculture has changed from family and cooperative operations to large, automated complexes. While productivity has increased as a result of the increased use of labor-saving machines, this rise has been directly dependent upon the rise of energy consumption. It has been estimated that the equivalent of 80 gallons of gasoline is

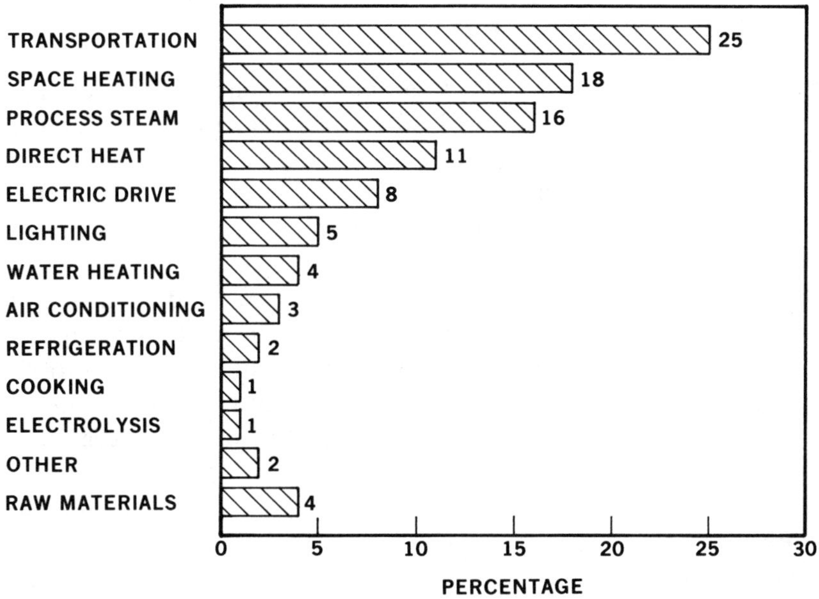

Figure 1-6 United States energy consumed by usage areas.[2]

presently required to grow a single acre of corn in the United States. Overall, the complete food chain in this country accounts for about one-seventh of the total energy use.[2]

Approximately 25% of the United States' energy consumption occurs in the transportation sector. Automobile drivers are by far the major consumers. Americans have traditionally been on the move. Henry Ford's production of the automobile and the discovery and utilization of cheap oil helped create a nation of people dependent upon mobility. There is now one automobile for every two men, women, and children in the country. The use of petroleum products has also increased. Heavier cars, pollution control measures, and inefficient, energy-extras have also contributed to this increased energy consumption. In an effort to reduce energy consumption from this sector, manufacturers have introduced smaller automobiles constructed of lighter materials. In addition, more efficient engine designs are now being considered.

Energy Statistics – The Future

What happens now? Do we, as a country, continue to expend our energy resources at the risk of losing our independence to foreign oil interests? Do we cut back and conserve to such an extent that zero energy growth is achieved and our life-styles are dramatically changed? Do we preserve our fossil resources for future generations? The answer, of course, lies between these extremes.

Energy growth will continue; however, more prudent use of energy resources must occur. Energy conservation must continue to be emphasized. Any projections of our future energy consumption are subject to speculation as to consumer energy needs, resource availability, unit energy costs, and the development of new energy related technologies.

In the years since our energy situation has become appallingly apparent, many major studies have attempted to project our relative growth rates and energy needs through the years 1985 and 2000.[11] Past energy situations have been identified and scrutinized down to the finest detail. Analysts are now able to explain how the present situation developed and what our historical patterns have been. Future extrapolations based on historical assumptions, however, cannot bridge the 1974–75 anomalies resulting from inflation, embargo, and economic events. Precisely these same factors that have altered our patterns of energy growth and consumption have also altered our techniques for forecasting.

Today's forecasters utilize sophisticated econometric supply–demand functions, based on a wide range of economic, social, and technological parameters. Even though this new forecasting base has been in existence only a short period of time and even though certain elements of the forecasting problem are beyond the state-of-the-art in forecasting, studies using this approach can yield meaningful projections of what the future holds. Figure 1-7 and Table 1-1 provide a survey of studies projecting energy growth rates to the year 2000. Variations occur and are significant. From Table 1-1, for total energy consumed, the range examined goes from approximately 93 quads (quadrillion Btu's) through 124 quads in 1985 and from 123 quads to 202 quads in 2000 A.D. Increased energy usage is projected in all major sectors.

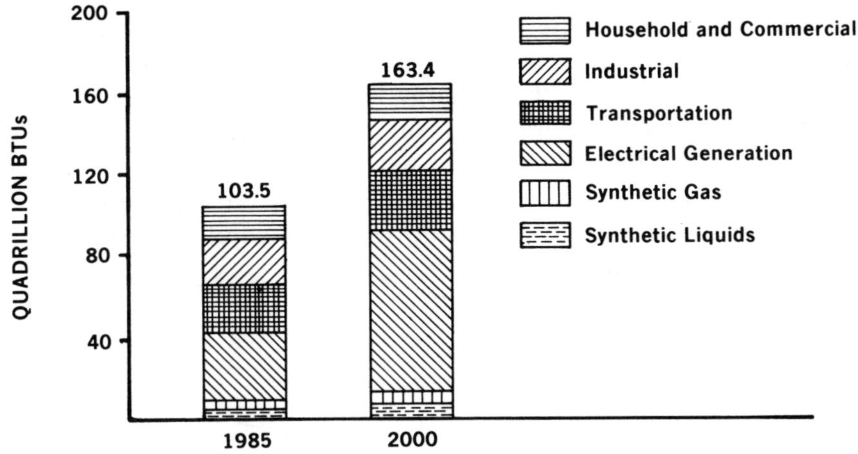

Figure 1-7 Average estimated United States energy consumption by major sectors through 2000.[3]

Table 1-1 Survey of Studies Projecting Total Energy Growth Rates to the Year 2000.[12]

	Growth Rate to 1985 (%)	Quads of Total Energy in 1985	Growth Rate to 2000 (%)	Quads of Total Energy in 2000
Chase Manhattan Study	5.0	124	-	-
Resources for the Future Study	-	-	3.5	192
Department of the Interior	3.7	118	3.5	192
Stanford Research Institute Study	3.8	119	-	-
Federal Power Commission Forecast	3.6	116	3.7	202
Council on Environmental Quality	-	-	1.8	123
Lawrence Livermore Laboratory Study	3.3	112	-	-
Atomic Energy Commission Study	3.7	118	3.6	197
National Academy of Engineering Study	2.9	107	-	-
Ford Foundation Study	1.7	93	1.9	126
FEA Project Independence Report	2.6	103	-	-

The energy consumption growth rate in the 1952 to 1972 period averaged 3.6% per year. Most recent forecasts for the future range from 1.3% to 3.1% growth in energy consumption per year. Using conservative estimates of growth in energy consumption, conservative predictions as to the availability of new energy technologies, and aggressive projections relative to natural gas and oil exploration, the distribution of energy sources for the years 1985 and 2000 should be essentially as displayed in Figure 1-8.

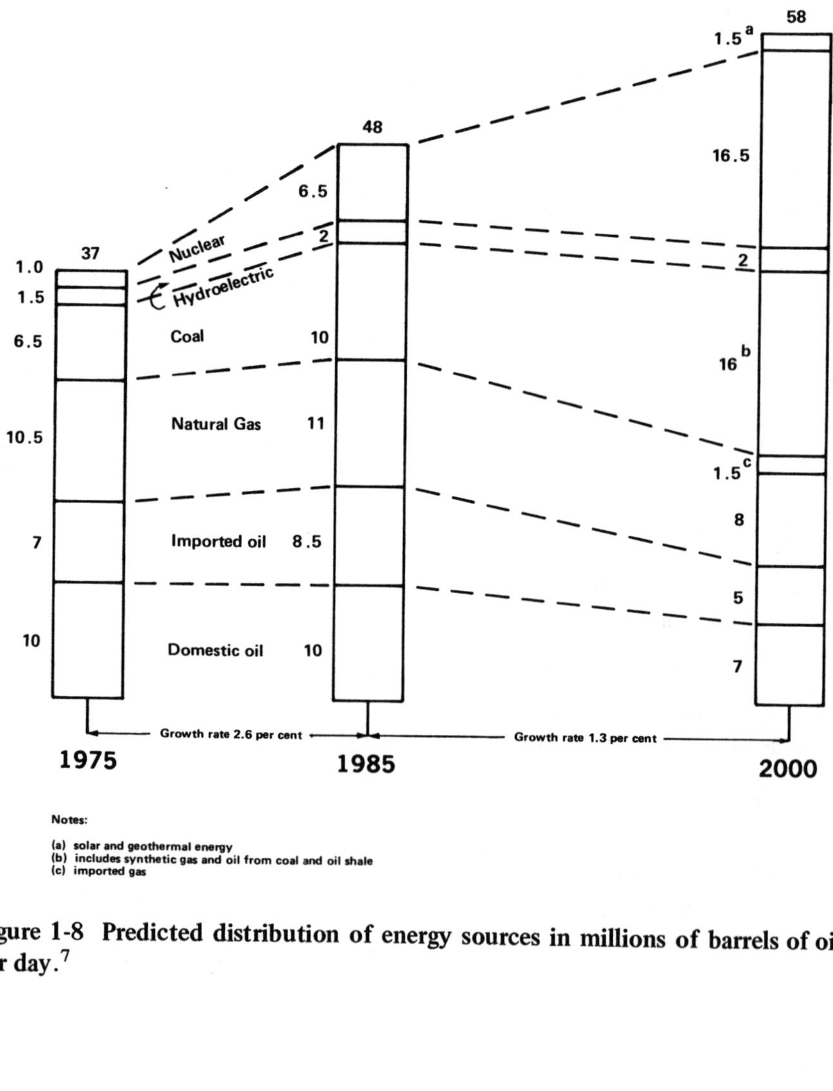

Figure 1-8 Predicted distribution of energy sources in millions of barrels of oil per day.[7]

Today, United States citizens represent 6% of the world's population but utilize 33% of the energy consumed in the entire world. To reduce energy demand as much as possible is a national goal. At the same time, increased employment, better production, and environmental quality are all assumed equally important. Some of the resources required to meet the energy demand of the future, in light of external societal constraints, are presented in the following chapters. (A list of state agencies that can be contacted for additional reference material is presented in Appendix II.)

Whether the future source of energy is nuclear or fossil fuel, synthetic or natural, new sites will be needed. A majority of these sites will be for plants that produce electricity. For this reason, the material which follows will emphasize the power plant siting problem within the context of meeting the nation's energy needs through the production of electricity.

ELECTRICITY

> The business of selling electricity to the public had an embryonic beginning in 1879 with California Electric Co., in San Francisco, but the full concept did not begin to flower until three years later, when Edison's Pearl Street Station started operation. Here, for the first time, the incandescent lamp and all the elements required for public supply were brought together. Only twenty years later, there were 3,620 electric utility systems in the United States.[13]

From this beginning the electric utility industry has grown to its present-day position of dominance. Although guided by distinct patterns of structure, the growth aspect has always remained a constant. The historic growth rate has averaged approximately 7% per year, as compared to the total energy growth rate of less than 4% (see Figure 1-9).[14] This growth rate resulted in a doubling of capacity every ten years.

The same situations that struck down the historic energy consumption growth, also affected the utilization of electricity. In 1974 many electric systems showed a decrease in growth as consumer conservation and a declining economy forced curtailments in energy usage. In 1975 there was a small increase in demand but far less than expected from historic trends. The consumption increase in 1976 over 1975 brought present rates of growth somewhat closer to past rates.[15]

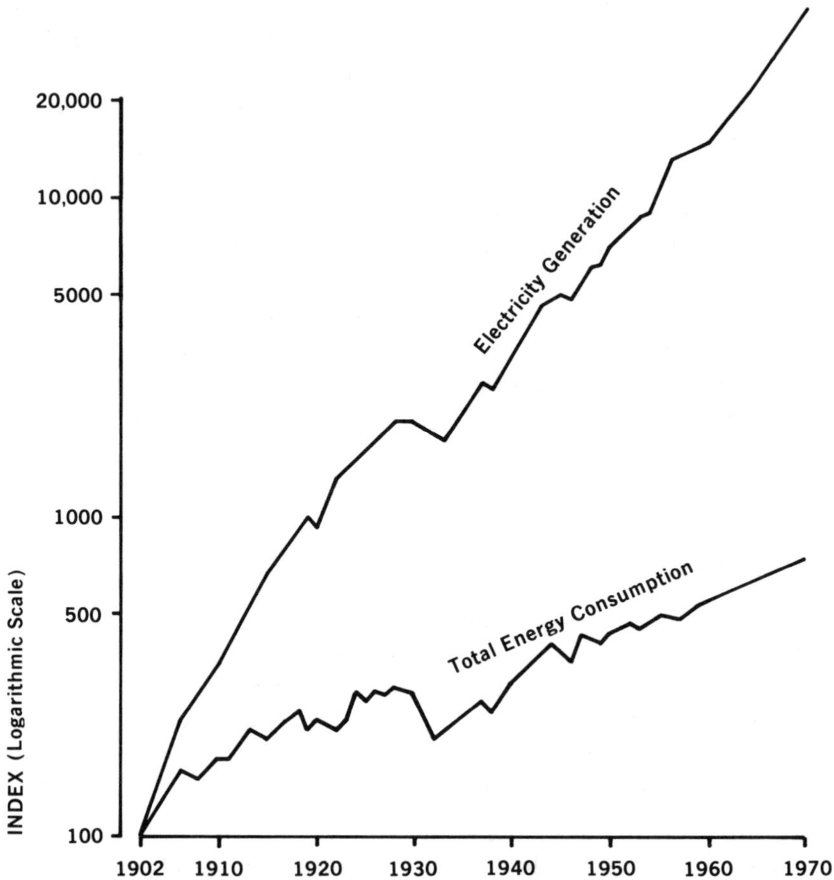

Figure 1-9 Comparison of electricity generation growth and total energy consumption.[14]

Because of the events of 1974 and 1975, both utility and regulatory forecasters are now forced to guess at future electrical needs. Projections of future electric demand range from 4.5% to 7.7% similar to the energy divergence noted earlier (see Table 1-2). Most studies agree, however, that the electrical energy form will remain the fastest growing energy form in terms of consumption. What is the rationale for this reasoning?

Table 1-2 Survey of Projected Electric Energy Consumption to the Year 2000.[12]

	Growth Rate to 1985 (%)	1985's Quads of Electric Consumption	Growth Rate to 2000 (%)	2000's Quads of Electric Consumption
Chase Manhattan Bank Study	7.5	4.4	-	-
Resources for the Future Study	-	-	6.8	10.8
Department of Interior Study	6.9	4.1	6.0	9.0
Stanford Research Institute Study	6.4	3.9	-	-
Federal Power Commission Forecasts	7.7	4.5	6.8	11.0
Council on Environmental Quality	-	-	4.5	6.0
Lawrence Livermore Laboratory Study	5.2	3.4	-	-
Atomic Energy Commission Study	6.6	4.0	6.5	10.1
National Academy of Engineering Study	7.0	-	-	-
Project Independence Report	6.6	4.0	-	-
Federal Power Commission Forecast	7.5	-	6.6	-

First, most projected new sources of energy will be designed to utilize existing technology, i.e., generation of electricity. New energy sources such as geothermal, wind, tidal, and solar will attempt to convert resources into electrical power. Conversion of fossil fuels such as coal, oil shale, and tar sands to synthetic fuels will most often be in conjunction with power generating facilities.

Second, industry will eventually convert to an electricity base. As natural gas becomes more expensive, it will lose its place as the major industrial source of energy. A recent study indicates that in the short run industry will switch to oil to meet their energy needs.[16] However, as this precious fuel becomes scarce it becomes more expensive and electricity will become the dominant choice. In many cases this electricity will be generated on-site, using synthetic fuels, garbage, and coal.

Third, a major reason for the future use of electricity is inertia. The electric utility industry has become an integral part of our society. It possesses a vast amount of inertia, a force that tends to continue movement in the same direction. Employment, investment, and production capability all enhance the use of electricity for fuel. The long lead times required to switch to alternate fuels versus the ready availability and flexibility of electric power will attract the consumer to electricity as opposed to raw primary fuels. Since additional electrical output will be required in the future, additional electric power plants will also be required.

NEED FOR POWER PLANT SITING

In January of 1975, President Gerald Ford noted in his State of the Union address that approximately 641 new electric generating plants must be in operation by 1985. To meet this need, he indicated that 200 new 1000 megawatt nuclear plants and 150 new 800 megawatt coal fired plants would be required over and above those units then being planned or constructed. In accompanying projections, the Federal Power Commission (FPC) and the Federal Energy Administration (FEA) indicated that in addition to 635 to 644 new generating units required between 1975 and 1985 (see Table 1-3) an additional 356 to 956 new generating units would be required shortly after 1985.

The need for additional power plant sites is obvious. While more than one unit can be built on a specific site, numerous sites will be needed. Because of the long construction period, plans for many of these sites are already underway. Typical project times are displayed in Figure 1-10. As noted, up to ten years are required to construct a power plant with the same time frame being required to construct associated facilities. It is important to note that the power plant sites

Table 1-3 Projected Requirements of Additional Electric Generating Facilities Based on FEA's Projected Need to 1985.[12]

	Additional Capacity (GWe)	Number of Units
Nuclear	184 to 220	184 to 220
Coal	160 to 124	200 to 155
Combustion turbine	(129) 129	(260) 260
Other	(12) 12	- -
Totals	485 to 485	644 to 635

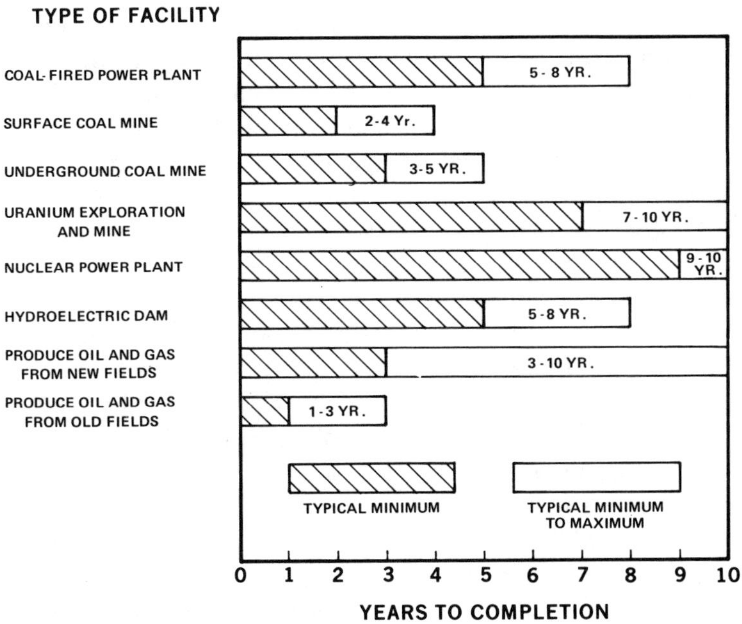

Figure 1-10 Typical project completion times.[4] (From go-ahead to production.)

are not only needed for new energy demands but also as replacement facilities for obsolete plants. For this reason, even if zero energy growth were to be achieved, new facilities and sites will continue to be required.

The type of generating plant that can be constructed on a given site is limited by economics, environmental requirements, and fuel availability. Each new plant, regardless of type, will make economic, environmental, and sociological impacts on the region surrounding the site. Because of these impacts, each new site and its accompanying facility will be exposed to an extensive review process; therefore, the function and rationale of the siting process must be understood.

ENERGY VS THE ENVIRONMENT—NEPA

On January 1, 1970, President Richard Nixon signed into law legislation that has had far-reaching consequences. This legislation was a result of the people's growing concern about environmental damage to lands and resources. While the movement is still evident today, most observers agree that it reached its peak in the late 1960s and early 1970s. In fact, all of the effort directed toward federal legislation, covered in more detail later, culminated in the passage of this act.

Specifically, this act is called the National Environmental Policy Act of 1969, or NEPA. Simply stated, the purpose of the act was (1) to encourage harmony between man and his environment, (2) to promote the prevention and elimination of environmental degradation, and (3) to enrich man's understanding of ecological systems and natural resources.[17] These steps were to be managed by "any agency of the federal government that is involved in proposals for legislation and other major federal actions significantly affecting the quality of human environment."[18]

As the law exists today, any power plant or associated facility which must apply for any permit from the federal government (whether it be an Environmental Protection Agency (EPA) permit to install, a U.S. Army Corps of Engineers discharge permit, or a Nuclear Regulatory Commission (NRC) permit to construct and operate) is under the jurisdiction of this act. An environmental assessment document, known as an environmental impact statement (EIS), is required for each addition, expansion, or new facility proposed by a utility. This document requires extensive data which serves to question the effort and the logic in the selection of the energy facility site.

With the passage of NEPA, the public can participate in utility decisions. Public input and testimony is accepted through environmental hearings held at the state level. In many states energy forecasts are published and distributed for comment. In many states hearings are held on these forecasts, giving the public an opportunity to express its opinion. What has been the actual effect, or impact, of this consumer involvement in the regulation of utility activities?

For the industry, almost overnight the burden and lead time for planning new sites has been effectively doubled and in some cases tripled. Legislative delays and red tape have increased to such an extent that utility planners are adding two to four years of lead time to projected construction schedules. Project costs are increasing similarly. In addition, "papyromania," the compulsive accumulation of papers, has been introduced to the regulatory and review system.

And yet, benefits have been reaped. A small section of the populace has been able to gain insight to and understanding of energy utility operation. To some extent, the populace as a whole has become more keenly aware of the complexity of the system supplying their energy needs. The detailed investigations and hearings of the siting review process have forced utilities to state explicitly their assumptions and methodologies. Often these thoughts are improved and refined through the give and take of the review process. This reorganization of thought and planning, although not major and far from complete, has helped the industry to better meet the expectations of the citizens it seeks to serve.

In effect, the public has now entered the power plant siting arena along with the utilities and regulators. Delay caused by such intervention is acceptable only if the consequences are understood by the public. To that end, more complete comprehension of the vast and complex iceberg of power plant siting, of which the rate-payers normally view only the tip, is critical. The chapters that follow will attempt to reveal more clearly the complexities of the power plant siting process.

REFERENCES

1. Wilson, R., "Some Perspectives on the Energy Question," *Nuclear News,* February 1976.

2. Simmons, H., "The Economics of America's Energy Future," U.S. Energy Research and Development Administration, 1975.

3. "United States Energy Through the Year 2000 (Revised)," Bureau of Mines, U.S. Department of Interior, December 1975.

4. "A National Plan for Energy Research and Development and Demonstration: Creating Energy Choices for the Future," U.S. Energy Research and Development Administration, April 1976.

5. Wilson, C., "A Plan for Energy Independence," *Foreign Affairs,* Vol. 51, No. 4, July 1973.

6. "1976 National Energy Outlook," U.S. Federal Energy Administration, U.S. Government Printing Office Document No. 0-204-986, 1976.

7. "Creating Energy Choices for the Future," U.S. Energy Research and Development Administration, 1975.

8. Gambs, G., "The Energy Crisis in the United States," Speech at the NAM Joint Policy Committee Conference, March 1973.

9. Nassikas, J., "National Energy Policy: Directions and Developments," *Proceedings of the American Power Conference,* May 1973.

10. Solberg, C., *Oil Power,* Mason/Carter, 1976.

11. Cook, J., "The Invisible Crisis," *Forbes,* July 15, 1976.

12. "The Need for Energy Facility Sites in the United States: 1975-1985 and 1985-2000," Environmental Policy Institute, June 1975.

13. Novick, S., "The Electric Power Industry," *Environment,* Vol. 17, No. 8, November 1975.

14. "The Electric Century 1874-1974," *Electrical World,* June 1, 1974.

15. "Monthly Energy Review," U.S. Federal Energy Administration, July, August, and September 1976.

16. "Potential Fuel Shifts," A Study by Stone and Webster Management Consultants, Inc., 1975.

17. "National Environmental Policy Act of 1969," Public Law 91-190, 91st Congress, S.1075, January 1, 1970, Title I, Section 102, Part C.

18. McCormick, J., "Content and Mechanics of Environmental Impact Statements," *Proceedings of the EIS Conference,* August 1973.

2

Power plant siting legislation

POWER PLANT SITING REGULATION

Prologue: The Electric Power Industry and the Environment

As briefly mentioned in Chapter 1, the United States' demand for electric power has been growing for a number of years. Most of this growth has been fostered by industrial expansion. Our country's socio-economic structure has insisted that power be provided from a continuing, reliable source. Because electric power system technology has been able to meet this demand, electricity will continue to be one of the dominant forces in our future energy supply picture.

Currently, fossil fuels (coal, oil, and gas) and nuclear fuels (uranium and plutonium) are the most common energy sources for electric power production. Most electricity is generated from coal, oil, or uranium. As a result, these fuel sources are the alternatives that the electric power industry must constantly compare. (A brief discussion of how electricity is generated is presented in Appendix III along with a brief description of various nuclear reactors.) Each fuel produces its own unique environmental impact when used in an electricity generating facility. The environmental movement of the late 1960s and early 1970s recognized and demanded that these impacts be enumerated, compared, and rated against some "acceptable" standards.

The first legislation enacted in the environmental field consisted of the 1965 Clean Air Act and the Water Quality Act of 1967. Both of these laws were later amended (by the Clean Air Act Amendments of 1970 and the Water Pollution Control Act Amendments of 1972) to broaden their scope of jurisdiction and to place them under the direct responsibility of a new federal agency now known as the Environmental Protection Agency (EPA). These laws were not directed at the electric power industry but at all industries which impacted the environment. However, this legislation, in essence, produced an enforcement structure that directed more detailed legislation specifically at the electric power industry.

The History of Electric Power Industry Regulation

The electric industry was born at the end of the nineteenth century. By the time Thomas Alva Edison evolved his "grand scheme" of selling electricity, all of the elements of an electric power system had already been invented. The following statement by Sheldon Novick details his point.

> The dynamo, the first commercially practical generator of electricity had been developed by Zenobe Gramme, a Belgian mechanic working in Paris. Electric lights, brilliant arc lamps used mainly for street lighting, were already in wide use in Europe and had been installed in several municipal lighting systems since the first such in San Francisco in 1879. The incandescent light (a carbon filament in an evacuated glass bulb) which would be needed for interior lighting, had been invented by Sir Joseph Swan, and perhaps by others, and the electric motor was available. What was lacking however, was a means of putting these elements together into a single workable system.[1]

Edison provided the catalyst to create this system by constructing and operating his Pearl Street central-generating station in 1882 and by selling "lighting" (electricity) to the financial district of New York.

As this new industry developed, Edison's young British secretary, Samuel Insull, became a dominant figure in the field. In fact, Insull is credited as being the motivation behind the electric power monopoly system that exists today. As individual urban systems grew, Insull envisioned a totally interconnected system having the capability to transmit power generated at any one point to many different areas. Insull felt that such an approach would prove economical and efficient for the hundreds of individual systems, each slaved to one distribution point. From his start with the Chicago Edison Company (now known as Commonwealth Edison), Insull, by mergers and acquisitions, gradually spread his domain over Illinois, the Midwest, and then over much of the country.

> By 1932, when this vast empire collapsed, Insull was president of 11 power companies, chairman of 65, and the director of 85. The actual extent of his control was never quite clear, but it seemed to at least some observers that he and J. P. Morgan, his great rival, controlled almost all of the nation's electric power business at the time of the Great Depression.[1]

Samuel Insull's companies along with other power companies across the nation emerged as natural monopolies. Interestingly enough, however, Insull was a prime mover in calling for public regulation of privately owned power monopolies similar to the regulation of the railroad industry. By 1898 regulation was viewed as the only answer to preserving private ownership of electric power companies.

In 1907 the first state power company statutes were established in Wisconsin and New York. These laws established a commission to set rates for privately owned power companies and to insure that stockholders were provided a fair rate of return on their investment. Each state in the union now possesses a similar regulation structure. Public utility commissions now set rates, hold hearings, and establish control over investor-owned utilities in much the same manner proposed by Samuel Insull but with the guidance of years of court decisions, precedent, and practice.

This second area for environmental legislation now becomes visible, namely regulation of power companies and their facilities. People in the late 1960s and early 1970s were aware that the electric power industry had been regulated for almost 70 years. Since this period had been marked by prices resulting from decreased costs of electrical generation, there had been no impetus to view the utilities as targets for concern and regulation as a tool for accomplishing environmental objectives. As power generation costs began to increase in the 1960s, these increases were passed through to the consumer, causing increased public scrutiny. Consequently, as other problems arose in the industry, public interest in further regulation flourished.

One sector of the public, the environmental movement, realized the potential that further industry regulation held as a means for stopping adverse environmental trends or, at least, mitigating their effects. With the regulatory visibility of the electric power industry and with the physical visibility of the power station complexes and their inherent environmental impacts, the environmental sector felt that selection of power plant sites could be more strictly controlled through the regulatory process. Hence, the concept of "power plant siting regulation" originated. As a result, many states enacted specific laws to regulate the effects that power plants and transmission lines could have upon a surrounding area.

The Philosophy of Power Plant Siting Regulations

Before moving to an overview of power plant siting legislation, a few parenthetical thoughts on power plant siting regulation should be considered. In regulating power plant siting one must keep in mind that the influence of a power system is not confined just to its geographic area of service. First, each individual system is interconnected to neighboring systems. Second, because of public pressure, new plants must often be constructed at locations far removed from their load centers. Third, construction costs and plant sizes often require that several utility companies build a single generation facility and utilize the power produced. Since a utility's sphere of influence extends beyond its geographic boundary of service, each siting action must be viewed in terms of its effect regionally and nationally.

Another consideration in the regulation of power plant siting is that some new facilities must be built to meet increased demand. Regulation, therefore, should be confined to the selection of sites and technologies that mitigate as many adverse impacts as possible. The regulation process should (1) involve a thorough review of all available options and sites so that related problems can be understood and studied and (2) culminate in a decision that clearly reflects the best long-term interests and needs of the region.

With this brief rationale behind the origins and philosophy of power siting regulation, the enacted legislation will be examined.

STATE LEGISLATION

In recent years power siting legislation has achieved expanding status at the state level. Figure 2-1 indicates how rapidly power plant siting

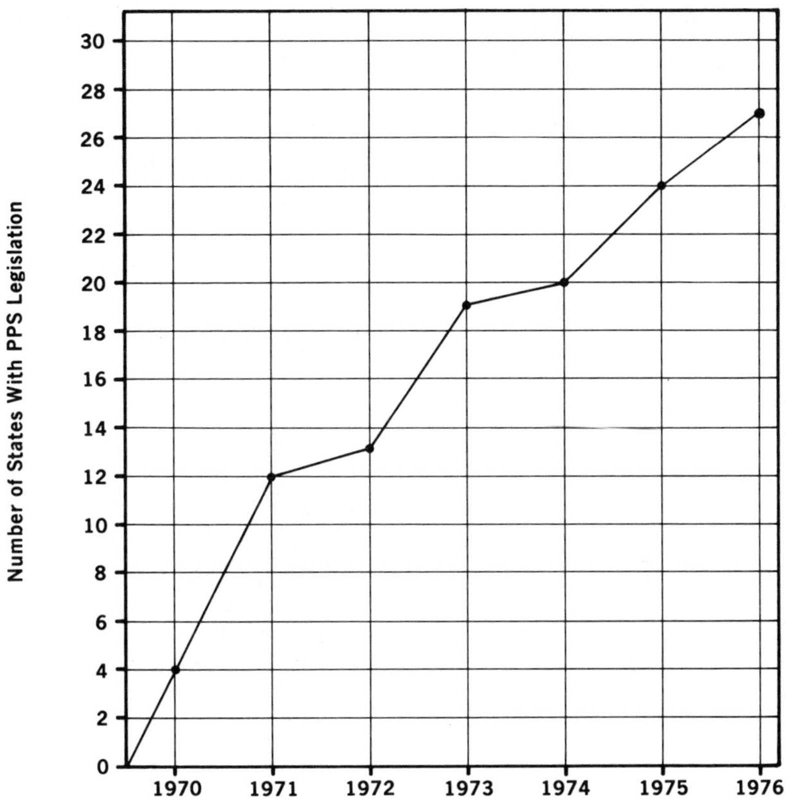

Figure 2-1 Evolution of state power plant siting legislation.

evolved as an integral function of state legislation. In 1970 only four states had enacted siting laws. As shown in Figure 2-1, the number of states possessing statutes relative to power siting has climbed to 27. Figure 2-2 displays those states that currently have agencies regulating the siting, operation, and maintenance of power facilities.

The first effect of the environmental movement was the involvement of numerous state and federal regulatory agencies in the siting process. Each agency was responsible for very narrow and independent review activities. Both environmentalists and the utility industry were overwhelmed with the increasing complexity and lead times resulting from the necessary hearings and procedures. The environmental movement pushed for a coordinated comprehensive review at state levels that was broader in scope and detail.

The industry, already plagued by labor problems, material delays, supply shortages, and increasing costs, now had to contend with increased "regulatory lag" or delay due to these lengthy environmental proceedings. As more agencies became involved in the review process, the lead time for government approval of facilities increased out of proportion. Each independent regulatory agency involved with the siting, construction, or initial operation of a facility, whether at the

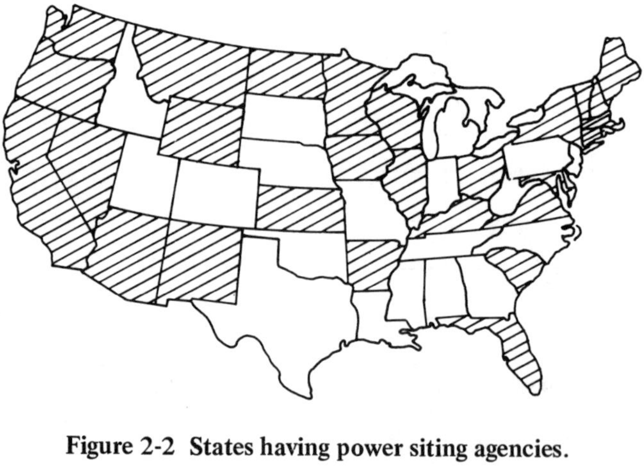

Figure 2-2 States having power siting agencies.

local, state, or federal level, could now impede a project. Because of this the utility industry also pushed for a more coordinated review.

Most of the early power siting legislation was prompted by the mutual concern of the environmental sector and the utility industry, and was based on several "model" statutes promulgated by various organizations and individuals. The most widely known and copied of these statutes was an act proposed by the National Association of Regulatory Utility Commissioners (NARUC) at their 1969 conference. This act is known as the Model State Utility Environmental Protection Act. (A copy is included in Appendix IV.) NARUC attempted to establish "one-stop" siting within the state government. *One-stop siting* is a concept whereby a single agency exerts control over all facets of governmental review and regulation of the siting, construction, and operation of a power plant or a transmission line.

With the passing of the initial legislation, unanticipated administrative and interpretive difficulties arose with the one-stop siting approach. First, a state agency, no matter how broad its statutory mandate, can not mandate control over a federal agency. (The areas of nuclear safety, air pollution, and water pollution were of primary concern here.) Second, as these "super agencies" were formed, problems arose with existing environmental and regulatory commissions. Those state agencies possessing strong authority from their own enabling statutes had difficulty yielding control on matters that they felt were still in their domain. Third, local authorities, township councils, and municipal governments found that their influence could only be effected through a state agency. Upset by loss of sovereignty, these governing bodies often refused to defer to state agencies.

As states adopted power siting commissions, the one-stop goal came to be regarded as a consolidation or coordination of all governmental review and regulatory functions (including the federal government). This action included public hearings available to all sides for additional input. Although much has been done to improve the original idea, controversy still remains over the merits of one-stop siting. An excellent comparative analysis of the various state programs is presented in Reference 2.

The diversity in power siting commission structures will now be examined. As of the end of 1976, the status of state power plant siting legislation could be separated into three major categories. As indicated in Tables 2-1, 2-2, and 2-3, state activity may be classified as minimal, moderate, or extensive.

Table 2-1 States Having Minimal Power Plant Siting Legislation.

Alabama	Michigan	South Dakota
Alaska	Mississippi	Tennessee
Colorado	Missouri	Texas
Delaware	North Carolina	Utah
Hawaii	Oklahoma	Virginia
Indiana	Pennsylvania	West Virginia
Louisiana	Rhode Island	

Table 2-2 States Having Moderate Power Plant Siting Legislation.

Arkansas	Kansas	New Jersey
Georgia	Kentucky	New Mexico
Idaho	Maine	North Dakota
Illinois	Nebraska	South Carolina
Iowa	Nevada	Vermont

Table 2-3 States Having Extensive Power Plant Siting Legislation.

Arizona	Massachusetts	Ohio
California	Minnesota	Oregon
Connecticut	Montana	Washington
Florida	New Hampshire	Wisconsin
Maryland	New York	Wyoming

State Legislation: Minimal

The type and extent of enacted power siting legislation has, in part, been dependent upon the amount of power generated in the jurisdictional area and upon the growth trends in that area. Those states with extensive power generation and corresponding high load growth patterns have usually enacted detailed legislation in the power siting field. In these cases the environmental impacts associated with power plants, combined with existing industrial pollution, tend to aggravate otherwise acceptable impacts in developed or developing areas. Thus, state industrial development is a major contributing factor to the presence of strong power siting legislation. Conversely, those states with controlled industrial development, extensive amounts of "undisturbed" acreage, and moderate need for power have chosen to control these effects by means other than power siting legislation.

The first category, as presented in Table 2-1, represents states that have never passed specific power siting legislation. Some of these states initiated power siting legislation but saw it defeated within the legislative process. As an alternative, many of these states have established mechanisms to investigate and analyze power siting impacts and control. For example, Alabama established an Energy Advisory Council; Colorado, an Energy Research Institute at the Colorado School of Mines; Louisiana, a Nuclear Energy Advisory Committee; Tennessee, a Joint House–Senate Task Force to study Power Plant Siting Legislation; Texas, a Governor's Advisory Committee on Power Plant Siting; and Pennsylvania, an Environmental Quality Board. Some of these efforts have resulted in proposed power plant siting statutes. (In fact both Colorado and Pennsylvania will likely pass such legislation in the near future.)

The legislative status of these states could change rapidly if industrial expansion and a growing need for power continue. Change will not be totally dependent on internal state conditions, however. To a certain extent, this change will depend upon exogenous influences to the states' jurisdiction. For example, one of the basic needs of a power plant is water (see Chapter 4). Thus, most large generating stations are sited on rivers or lakes. If the body of water chosen is a state boundary, coordination with the bordering state could be required.

In addition to the site-specific jurisdictional questions under review in these states, there are numerous ongoing general studies to determine power siting rationale and development within drainage basins of both rivers and lakes. The Ohio River Basin, the Mississippi River Basin, the Missouri River Basin, and the Great Lakes area are all undergoing analysis of one type or another associated with land management options. One objective of these studies is to unite involved states with a common goal in utilizing available resources to the fullest extent possible. Thus, as more attention is directed toward state cooperation in this field, some states may be forced to adopt power siting legislation in order to organize and detail their own interests in power siting jurisdiction.

State Legislation: Moderate

Attention is now focused on those states that have "moderate" power siting legislation. Table 2-2 lists the states placed in this category. Various types of organizational patterns and varying degrees of control are found among these states, but all possess a common concern over the impacts associated with the siting of large generating stations. All states have approved specific power plant siting legislation. These states can generally be grouped into three subcategories, namely states using a Public Service Commission, states using the Coastal Zone Management concept, and states using their environmental agencies.

Some of the states have given control of the review procedures to the Public Service Commission already existing within their state government. (Example states are Arkansas, Nevada, New Mexico, and Vermont.) In these jurisdictions the commission determines the public convenience, necessity, and environmental compatability of the proposed facility. A public hearing may be required in the process; however, the utility must obtain a certificate from the commission before it can begin to construct its plant. In most cases a detailed environmental review of the plant is neither required nor performed. Existing studies, literature searches, and utility-funded monitoring programs are accepted in lieu of independent, expensive environmental studies.

Also lacking in most of these jurisdictions is a required forecast of future load growth and plant additions. (Most states in the "extensive" category require utilities to file such a forecast.) The lack of this forecast prohibits the review body from reviewing the requested expansion in terms of long-range need.

The second type of jurisdiction placed under the moderate category consists of states with Coastal Zone Management legislation. (An example is South Carolina.) These states are concerned that new industrial developments not deteriorate the coastal region adjacent to the state's boundaries. Laws were enacted to control all development and land use in the coastal area; however, energy facilities have become the major jurisdictional targets. The necessity for large amounts of water for cooling, the physical location of fuel deposits, and the separation factor for nuclear plants, in some cases have prompted the use of these coastal regions for energy facility siting. Accordingly, states which feel that they have a mandate to maintain the status quo of their coastal zone have passed legislation to implement coastal zone concepts.

The third type of jurisdiction assigns responsibility to an environmental agency in the state government. That is, control of the review process is placed within an agency such as the Department of Natural Resources and Conservation, the State Health Department, or the Environmental Protection Office. (Example states in this grouping are Maine and Kentucky.) In addition to possessing problems similar to those found in states using a Public Service Commission, placement of the review activity in an environmental agency can result in specific nonenvironmental problems being overlooked. While general environmental aspects will normally be closely reviewed, many technological constraints and aggregated impacts will be overlooked or traded off in favor of preserving the environmental area.

State Legislation: Extensive

States with aggressive power siting programs have chosen a variety of ways to handle the review and decision-making process. Some of the more interesting programs are discussed in Chapter 3. For the present, however, this grouping of states, as enumerated in Table 2-3, will be viewed through three perspectives; one-stop siting, two-stop siting, and coordinated siting.

The "one-stop" siting agency evolved as a method of insuring total review of the proposed facility and its associated impacts. While a general evaluation of the one-stop idea was introduced earlier, some additional comments may aid in understanding the process.

Two major criteria used in evaluating effectiveness of siting programs are (1) timeliness, the ability of the process to complete its review in a reasonable time frame, and (2) balance, the fairness and comprehensiveness of the process.[3] Judgment in the former area is based upon the potential for delay through the administrative and judicial process. Depending upon the amount and type of public involvement and upon the control of the one-stop agency, the certification process could take longer with a one-stop agency than it did prior to the existence of such a body. Of course other factors can affect the time frame and could be included. Also, delay problems are not always the fault of the system as will be discussed in Chapter 3. But the potential for delay must be a matter for concern.

Although delays will not always be adverse, extensive delays in implementing an expansion program can result in escalated final cost. This cost is then reflected in the consumer's energy cost. Small delays can prove fortuitous when the time is devoted to investigating a forgotten design detail or to reviewing an unnecessary impact that can be easily mitigated.

As state organizations are drawn into the one-stop review process, the multifaceted commission coordinating the review will tend to reflect individual agency interests in the final decision. However, the degree to which these agencies are involved is directly related to the impact of their comments on the process. Hence, balance comes into play. For example, if an environmental commission submits more comments and demands more information for its review than the Public Service Commission, the final decision will, in all likelihood, reflect more stringent environmental controls at the expense of productivity and favorable rates. Conversely, lack of adequate environmental review could result in a plant being constructed without adequate controls thereby requiring expensive retrofitting after operation in order to meet regulations.

The states in the one-stop group (Arizona, California, Connecticut, Maryland, Massachusetts, Minnesota, North Dakota, Ohio, Oregon, Washington, and Wyoming) have encountered problems relative to rendering timely, balanced reviews. Fortunately, the statutory authority of most of these states was patterned after the NARUC model. As a result, solutions have been effected by a simple Rules and Regulations change. These changes are written and promulgated apart from the law; thus, modification is made without legislative review.

One legislative variation that has produced an interesting slant on the review process is the "two-stop" agency, the second type of aggressive power plant siting legislation. States with this type of organization usually divide review responsibility between the Public Service

Commission and an environmental commission. This approach bypasses the necessity for establishing a new authority that usually creates attendant problems associated with organization and implementation. The existing commissions and staff are used. Complementing review procedures are used to insure full and adequate review before certification. Balance is achieved since two central agencies operating from a strong responsibility base must resolve conflicts through compromise.

The final approach discussed in this section concerns those states that possess siting laws, either one-stop or other, that have structured the review process into a coordinated procedure. The coordinated procedure utilizes a small in-house central staff which draws upon the expertise and experience of the staffs of the member commissions. If handled properly, final decisions involve all relevant points of view with interagency difficulties worked out in the process.

In this process total preemption of regulating commissions by a super-agency is avoided. Rather, selective preemption is exercised, with full knowledge and agreement among the member commissions and interested parties. Thus, compliance with specific statutes, outside of the power siting statute, can be mandated within the commission's decision. In addition, sufficient expertise is available for review and testimony to insure proper discussion of the often complex issues.

In comparison with the various state approaches to the situation, the federal approach is now inspected, and a historical perspective is provided.

FEDERAL LEGISLATION

Federal legislation dealing specifically with utility systems has been introduced in Washington many times. In the mid-60s after the 1965 Northeast power blackout, bills were introduced relating to system reliability improvements. At the same time Reliability Councils were established through the Federal Power Commission (FPC). These councils (Figure 2-3) were formed to insure an adequate transmission line interconnection and to coordinate systems expansion so that a recurrence of the 1965 disaster could be avoided.

In the late 1960s, when public emphasis changed to environmental protection, the federal impetus also changed. Numerous federal

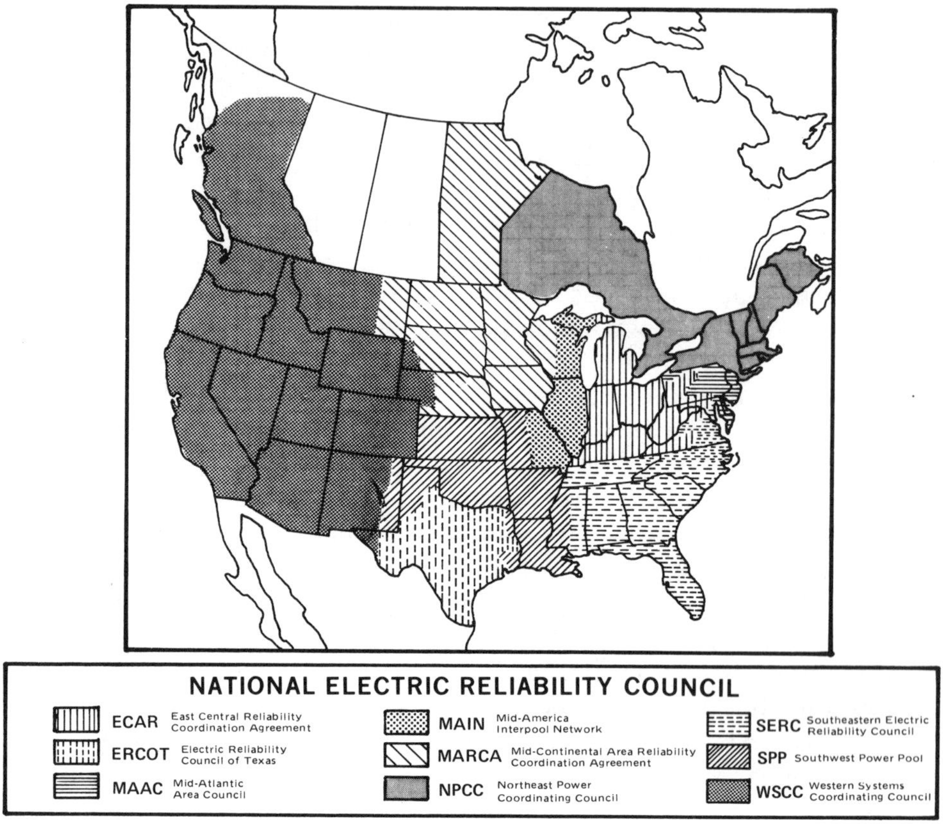

Figure 2-3 Regional electric reliability councils.

statutes were promulgated relating to general environmental protection. None of these statutes, however, were directed specifically at power plants. Many power plant siting bills were proposed. Most of these bills vested authority with a federal agency as was done in NEPA. In 1972, for example, seventeen bills were studied by the United States House of Representatives Subcommittee on Communications and Power of the Committee on Interstate and Foreign Commerce. None of these bills were passed.

In the years following 1972 the emphasis shifted again. The new emphasis was stimulated by the Arab oil embargo of 1973. Energy legislation, combined with concern over nuclear safety and licensing problems, evolved a new set of power plant siting bills. At the date of this writing, no bill has been passed by Congress.

An important alternative approach to direct legislative involvement exists at the federal level. This program, known as the State/Federal Power Plant Siting Program, is an integral step in the development of a uniform national policy on power plant siting. Although this program deals mostly with nuclear programs and must be approached through the Nuclear Regulatory Commission Office of State Programs, all aspects of power siting are discussed in the forum. The federal involvement in this program includes agencies such as the Federal Energy Office, the Federal Power Commission, the Energy Research and Development Administration, and the Environmental Protection Agency. Annual meetings have been held since the first state/federal conference in 1975. As a result of the first conference, several useful innovations were adopted by the federal commissions. For example, closer coordination and joint hearings for nuclear power plant certification are now being implemented. Closer state contact is also being maintained by ongoing task forces. Nuclear energy centers and associated impacts are being investigated. These actions have aided the states in organizing and implementing their individual programs and have provided an avenue for state input at the federal level.

CONCLUSION

From its conception, the utility industry has been extremely visible and susceptible to public control and criticism. Control has been exercised at both the state and federal levels. This control has evolved over a period of years through a complicated, often misunderstood, set of governing decisions, laws, and situations. The authority and impetus of the legislation has been directed by constituent concern which has changed over the years. In the early 1900s fear of monopoly evoked the establishment of Public Service Commissions to control monopoly utilities and set rates. In the 1960s the public demanded strict control over industrial pollution and called for environmental regulation of utilities.

In recent years, the public outlook has again changed. Now energy is the concern of the public. This concern affects utilities in a different and more subtle manner than encountered previously. Faced with a changing power demand and with an effective loss of one year of growth as a result of the 1973 oil crisis, the utilities must forecast their load growth more carefully than before. Public concern for accurate information has led to some states requiring the submission of load forecasts to substantiate the need for future plants and facilities. The *need* for the power plants has now become a major issue.

Utilities continue to stress the minimal problems associated with building "excess" capacity as compared to the imminent danger of lack of capacity. They contend that lack of capacity will result in unreliable system operations and outages. Consumers, on the other hand, are intensely concerned about ever-increasing costs, whether these costs result from increased fuel costs, operating costs, construction costs, or excess capacity. In particular, the public is concerned that poor managerial planning and improper utilization of resources not force the cost of electricity upward.

With this, pressure load forecasts are now closely scrutinized along with system planning and management costing. In many instances this scrutiny has a positive effect in that utilities have been forced to reevaluate their planning methodologies. Public Service Commissions across the nation are now having utility management audits conducted. The results of these studies are often used in public hearings on power plant considerations. Thus power plant siting has now assumed a prominent status in the public eye.

One-stop agencies, implemented as a novel idea in the early 1970s, have now been beset by bureaucratic problems. The effectiveness of these programs is now being measured and analyzed. Two of the measurement criteria are timeliness and balance. Two-stop agencies and coordinative agencies are now being touted as alternatives that possess the capability of overcoming the problems of one-stop siting. At the same time, more expansive energy commissions are being created at the state level. These state organizational structures deal with the entire energy field and associated impacts and predictions. They authorize research expenditures and govern construction of facilities. Therefore, it appears that this broader approach to energy facility siting is the wave of the future.

Within the federal framework, no significant power siting legislation has yet been passed. Given the site-specific nature of power plant siting and given the difficulties that have arisen in implementing state legislation, the viability or effectiveness of a national power plant siting bill is questioned. The most important action that federal commissions can undertake is continued involvement in state programs. Advice and input should be offered through such forums as the State-Federal Siting Program, the National Governors' Conference, and the various regional advisory boards. An important federal input to state activity exists through the various basin studies and land-use management studies on the coastal zones. This input should be encouraged so long as it guides rather than forces regional actions.

Finally, it appears that public involvement in the power plant siting process, while changing in thrust, is still increasing. As new aspects of power siting surface, more involvement is expected. The next chapter will attempt to relate the myriad of involved commissions, at all levels, to public involvement and understanding. This understanding is necessary if the public is to maintain its desired control over such regulatory bodies and contribute significantly to the process.

REFERENCES

1. Novick, S., "The Electric Power Industry," *Environmental Magazine,* Vol. 17, No. 8, November 1975.
2. Felman, D., et al, "CAPS: Comparative Analysis of Power Siting in the U.S.," Ohio Power Siting Commission, Columbus, Ohio, June 1, 1977.
3. Winterland, V., "One Stop Siting for Major Utility Facilities: Intent and Implementation," Policy Paper, Ohio State University, 1976.

3

Regulatory agencies

INTRODUCTION

As noted in the previous chapter, the public, with an increasing awareness of the issues surrounding electric energy production and sales, has started to demand governmental action. The legislative response to date to these demands has been a complicated, interwoven, sometimes conflicting system of laws at the various levels of government. The net result of this legislation has been the development of regulatory agencies that have been given broad authority to interpret their enabling statutes. While these public agencies may be somewhat guided by years of court decisions and precedent, the authority to promulgate rules and regulations has resulted in a second set of complex requirements which the electric utility industry must meet in order to gain approval for construction of new energy facilities.

The National Environmental Policy Act places the responsibility for the preparation of environmental assessments with federal agencies. Any plant construction that in any way involves a federal agency must comply with the requirements of this act, as explained in Chapter 1. While the requirements of this act will undoubtedly affect all future power plant development, there are numerous other agencies which will also be involved. Throughout the entire gamut of preconstruction planning and approval, and continuing through construction and operation phases, a utility must develop contacts with agencies at all levels of government. In fact, each agency having an interest in power projects within its jurisdiction and possessing some approval or regulatory authority must be approached.

Utilities have been forced to evolve a new form of interaction with and awareness of public officials at all levels. The activity of obtaining site selection approval has become a complex enterprise. A trend, however, is developing where all discussions involving utility and public officials are publicly announced and held in open forums, accessible to any interested citizens.

While these types of contacts and requirements are becoming familiar to the utilities, it is important that the public also understand the methods and rationale of the approval process. It would require another entire book to explain in detail the workings of each regulatory agency involved in the power siting process, but in this chapter an outline of the various levels of activity and a brief discussion of the agencies involved in the siting process are given. Since the involvement of regulatory authority has been blamed for drastically increasing the lead time for planning and constructing new power plants, discussions of some of the more innovative approaches to siting at the state regulatory level and of some of the delay factors currently presented in many siting processes are also included.

LOCAL REGULATORY AGENCIES

Local agencies are those existing in municipalities, counties, and townships or through regional organizations. Typical agencies involved in the power plant siting process are listed in Table 3-1. These agencies generally contain approval authority for any type of major construction project within municipal, county, or township jurisdiction.

Table 3-1 Local Regulatory Agencies.

Municipal
City Council or Board of Alderman
City Planning Commission
City Air Quality Board
City Water Quality Board
City Health Department
City Aviation Board
City Department of Public Works
City Zoning Board
County/Township/Parish
County Commissioners/Township Trustees
County Planning Commission
County Air Quality Board
County Water Quality Board
County Health Department
County Engineer
Civil Defense Office
Water District
Port Authority

Requirements may vary greatly from local government to local government. However, initial approval must normally be obtained from these authorities before proceeding to state or federal agencies. Adherence to local statute is required, and local approval is normally vested in these organizations.

These agencies must not only authorize specific power plant projects but, in many cases, must also be responsible for aiding the local community in mitigating social impacts (discussed in Chapter 5) associated with power plant construction and operation. Collection and disbursement of additional tax revenue, planning and funding of new schools and public services, and the effects of accelerated growth must also be considered by these agencies.

The fragmentation of responsibility at the local level often complicates public involvement in the siting process. Inexperience with power plant siting by public officials often frustrates these officials, hampering meaningful public interaction. The following sections describe the relationship of a few of the local agencies to the siting process and how the public can relate to these agencies.

Air, Water, and Health Boards

These agencies are responsible for (1) certifying that industrial health, safety, and environmental criteria are satisfied and (2) issuing specific approval, by means of permits, for the operation of individual industrial facilities. Power plant permits are generally issued with a minimum of delay provided all operating parameters satisfy agency criteria. These is little opportunity for direct public involvement unless the permit process involves a public hearing. In these situations, constructive public input can be assured only if there is a familiarity with local quality criteria.

Zoning Boards, Port Authorities, Engineering, and Public Works

These agencies are responsible for (1) certifying that public convenience criteria are satisfied, (2) approving patterns of public activity in the area, and/or (3) recommending a course of action to elected officials who will validate patterns of public activity. For example, land-use changes, such as rezoning or port authority authorization, result from favorable engineering and public work analyses, favorable zoning board or port authority hearings, and/or positive approval by elected officials. While the criteria in the area of public convenience may be well defined, decisions relative to the application of these criteria are open to interpretation by a number of public officials. As a result, several points of interaction are available to the public. In each case, however, interaction must specifically relate to the domain of responsibility of that agency.

Councils, Commissions, and Trustees

These officials, their staffs, and their support agencies play important roles in the power plant siting process. Activities range from soliciting new community capital investments to legislative resolution of issues prohibiting these investments to mitigating adverse impacts resulting from these investments. These activities are not mandated by explicit legislative authorization but result from public pressure for continued economic growth in that geographic region. Public interaction with these officials must therefore be limited to economic issues, public acceptability, and, for elected officials, voter response. It is important to remember that initial utility contact in a community usually occurs in this sector of local government.

STATE REGULATORY AGENCIES

Although local agencies are most directly affected by the siting process, state regulatory agencies often have more control and direct legislative mandate over utility operation. Some typical state agencies are listed in Table 3-2. In many states, a significant number of these agencies must submit individual approval for siting. In a few states, siting approval is consolidated through a specific program. This section will first address the common state interests in several specific areas and then review a few specific state regulatory programs for power plant siting.

Air, Water, Solid Waste

A Division of Air Pollution is normally present within state government. The authority for state control of air emissions originated with the Clear Air Act of 1965. Normally, this agency will enforce both the state's air quality standards and the air quality standards propagated by the U.S. Environmental Protection Agency.

The Water Quality Act of 1967 requires any industry planning to make any discharge into the waters of a state to submit an application for a permit to a state Division of Water Resources or its equivalent. In the case of a power plant, a utility must include a detailed description of the proposed plant. This includes proposed uses of the water, the sources and characteristics of the discharge, descriptions of the discharge facilities, and the discharge volume and quality parameters. A description must also be given of the treatment methods investigated and the treatment method selected, including considerations given to segregating certain wastes or to reducing in-plant waste. The application must describe the impact the effluent will have on the receiving waters.

Table 3-2 State Regulatory Agencies.

Aeronautics Commission
Archaeological Commission
Historical Commission
Attorney General
Department of Agriculture
Department of Conservation
Department of Commerce
Department of Environmental Protection
Department of Finance and Administration
Department of Fish and Wildlife
Department of Human Resources
Department of Mines and Minerals
Department of Natural Resources
Department of Parks and Recreation
Department of Transportation
Division of Air Pollution
Division of Forestry
Division of Solid Waste
Division of Water Control
Office of Local Government
Office of Planning and Research

The federal legislative process produced the Energy Conservation and Resource Act of 1976 to regulate solid-waste disposal methods, amounts, and types. Most states contain a Division of Solid Waste, or equivalent, that oversees industry compliance with these federal standards and with pertinent state standards.

In many states, these three functions are organized into one process. Some states require public hearings before permits can be issued, providing an excellent forum for public input. In these cases, however, constructive input requires a detailed knowledge of the state's quality standards and how the proposed facility's design will relate to these standards.

Conservation, Fish and Wildlife, Forestry

These state agencies deal with the preservation and proper utilization of a state's natural resources. These agencies can prevent power plant siting in specific areas such as along wild and scenic rivers, set aside for preservation. These agencies can also be involved in the site approval process such as when a power plant will have a lake or reservoir that will be used as a recreational facility.

The fish and wildlife agency can be a point of important public interaction. This agency possesses the sensitive responsibility of maintaining the environmental quality of the fauna. Since adverse construction and operational effects on rare and endangered species must be avoided, extensive reviews are generally conducted by this agency before any permit is issued. The study reports of the fish and wildlife agency can be a valuable source of public information regarding site effects on animal life cycles.

Other Agencies

While the state agencies referenced may possess the most significant voice in the siting process, other agencies can definitely influence the success or failure of a particular siting effort. The Public Service Commission must approve the type and amount of financing for the plant. The Department of Human Resources must certify that the labor market will be adequate to meet the needs of the facility. When tall stacks or cooling towers are required, the state's Aeronautics Commission must approve clearances for flight paths. Archaeological and Historical Commissions must ensure preservation of the state's heritage and prevent siting activities from impacting significant landmarks. The Department of Transportation must ensure that transportation facilities are adequate for plant construction and operation. Each of these agencies provides a point of public input as it relates to the specific areas of responsibility of the agency.

Specific State Regulatory Programs

Because of regulation complexity at the state level, some states as noted in Chapter 2, have enacted legislation that consolidates many of the state siting activities. Several of these programs will now be reviewed.

Maryland.[1,2] The Maryland Power Plant Siting Act of 1971 created a Power Plant Siting Program (PPSP) within the Department of Natural Resources. The statute included several unique features in the program, such as (1) state authority to acquire and hold in reserve sites suitable for future plant construction, (2) establishment of a broad continuing power plant environmental research program to support site evaluation and selection and related environmental and land-use considerations, (3) sharing of responsibilities between the state and public utilities for long-range planning, and (4) an Environmental Trust Fund supported by a surcharge on all electricity sales (ranging between 0.1 and 0.3 mil per kilowatt-hour) to provide funds for environmental review of proposed sites, siting research activities, and purchase of potential sites.

PPSP is a research and advisory program, not a regulatory program. It addresses trade-offs between environmental, economic, engineering, and scheduling interests. The Public Service Commission rules on site and facility suitability. Because the state establishes and purchases a site inventory, the Maryland program is considered to be among the most innovative in operation today.

Washington.[3, 4] In 1970, the Washington State Thermal Power Plant Siting Act was enacted as the nation's first one-stop siting law. The act created an Energy Facility Site Evaluation Council with the power to adopt rules and regulations and to develop criteria for the design and location of thermal power plants and other energy facilities.

The Council consists of representatives from the various state agencies and commissions having an interest in the siting process. The Council attempts to develop sensible trade-offs where agency responsibilities conflict and to arrive at meaningful compromises when diverse interests are at variance. The one-stop forum allows singular preconstruction certification, technical review, planning, and public hearings.

Strict interpretation of the enabling statute would exclude all local agencies from the siting process. However, in this state, most local authorities have not surrendered their autonomy.

California.[1, 2, 5–7] On January 7, 1975, the California Energy Resources Conservation and Development Commission (ERCDC) was formed. The ERCDC consists of five appointed members and two ex officio members, namely the Secretary of the Resource Agency and the President of the Public Utilities Commission. This commission is responsible for (1) compiling all standards to be met in designing, siting, and operating facilities in the state, (2) formally approving or disapproving applications for constructing energy facilities within the state, (3) preparing integrated plans specifying actions to be taken in the event of an emergency, and (4) conducting an ongoing assessment of the opportunities and constraints presented by all energy forms including activities necessary to develop proper energy conservation practices. The ERCDC's role is one of planning, forecasting, certifying new plant sites, and actively utilizing all forms of energy within the state.

Ohio.[1, 2, 8] The Ohio Power Siting Commission was established in 1972 and became effective in 1974. This agency is composed of five members consisting of the Chairman of the Ohio Environmental Protection Agency, the Chairman of the Public Utilities Commission, the heads of the Department of Economics and Community Development and the Department of Health, and a citizen member. The citizen member must be an engineer appointed by the Governor.

Ohio has a one-stop law; however, the statutory mandate for approving construction of new power plants and transmission lines is conducted in a coordinative manner. As a result, the Commission's main function becomes one of fostering full and complete cooperation between all state agencies responsible for inputs in the siting process. While the Commission possesses direct preemptive power over local authorities through the public hearing process and through a program of state coordinators, this preemptive process has been seldom used.

Florida.[1, 2, 9] The Florida Siting Program was organized on July 1, 1973, and reorganized on July 1, 1975. While the program is considered to be a one-stop program, it is an excellent example of a two-stop process. A prime advantage of the Florida Siting Program is that it avoids many of the problems evidenced in the normal one-stop approach.

In Florida's program the Environmental Regulation Commission is responsible for reviewing applications to certify a site from the various environmental perspectives. Simultaneously, the Public Service Commission determines the need for the plant – the other major component of the siting decision. This review relative to need utilizes long-term forecasts submitted by the utilities to the Division of State Planning. The Governor and his cabinet act as the final authority on the application by approving or denying the issuance of a certificate. If approved, the certificate binds the state and all state agencies to the approval of the site and to the construction and operation of the proposed power plant and its related facilities.

Other State Programs.[10–23] Each state program contains its own unique features. Variations in the one-stop process may be found in the siting programs of Arizona, Connecticut, New York, and Oregon. Variations in the two-stop process may be found in the siting programs of Arkansas, Maine, and Wisconsin. States with relatively new laws include Iowa, Kansas, Montana, and Wyoming.

FEDERAL REGULATORY AGENCIES

Under the requirements of NEPA, the federal government is responsible for the development of environmental impact statements. There are many federal agencies that have a potential for involvement. A representative list is given in Table 3-3. While most of these agencies could be involved in any particular licensing operation, five of them play a significant role in the siting process.

Environmental Protection Agency

The Environmental Protection Agency (EPA) was established on December 2, 1970, by Reorganization Plan Number 3 of 1970 which provided for the consolidation of pollution control and abatement responsibilities from all or part of the following agencies: the Department of the Interior; the Department of Health, Education, and Welfare; the Department of Agriculture; the Atomic Energy Commission; the Council on Environmental Quality; and the Federal Radiation Council. The EPA initially was to set standards and criteria for the protection of the existing environment.

The Clean Air Act Amendments of 1970, the Water Pollution Control Act Amendments of 1972, and other legislative proceedings broadened the authority of the EPA to cover many areas. These areas now include jurisdiction in ambient air quality, water quality, radiation protection, public drinking water, solid waste management, sources of specific pollution, and discharge into navigable waters and their tributaries. Local offices of the EPA, supported by field

Table 3-3 Federal Regulatory Agencies.*

Corps of Engineers, U.S. Army
Council on Environmental Quality
Department of Agriculture
• Forest Service
Department of Commerce
• National Oceanographic and Atmospheric Agency
• Coastal Zone Management Program
Department of Defense
Department of the Interior
• Bureau of Indian Affairs
• Fish and Wildlife Service
• National Park Service
Department of Housing and Urban Development
Department of Transportation
Energy Resource Development Administration
Environmental Protection Agency
Federal Aviation Administration
Federal Energy Administration
Federal Power Commission
Nuclear Regulatory Commission
Securities and Exchange Commission

*This table reflects federal regulatory agencies prior to the establishment of the Department of Energy in 1977. As this department becomes operational, certain agency realignments will be effected changing the structure of this table.

organizations in various states, perform surveillance and inspection to determine compliance with established environmental standards, pollution source standards, permits, and product registrations. In general, while states are expected to exercise primary responsibility for compliance and enforcement, federal action is taken where this activity is not performed.

The Federal Water Pollution Control Act Amendments extended the EPA's authority to all navigable waters. These amendments also established two rational goals, namely that discharge of pollutants into navigable water be eliminated by 1985 and that, wherever attainable, by July 1, 1983, water quality be brought to a point that fish, shellfish, and wildlife are protected and recreation can occur in and on the water. To accomplish these goals, a regulation program was established based upon issuance of permits under the National Pollutant Discharge Elimination System (NPDES) permit program (Section 402 of the Federal Water Pollution Control Act). Those states that meet requirements set by the EPA can administer the NPDES program and issue permits to dischargers, including power plants, along guidelines established by EPA and subject to EPA review.

Federal Power Commission

The Federal Power Act of 1920 assigned to the Federal Power Commission (FPC) the responsibility of approving the sites of all new hydroelectric generating facilities. This legislation (now known as Part I of the act) provided for a comprehensive program of national regulation for hydroelectric plants as opposed to the case-by-case procedure that resulted from the Rivers and Harbors Act of 1899. In 1935, Congress enacted Parts II and III of the Federal Power Act giving the Commission authority in the areas of power planning and accounting.

The FPC licenses the construction and operation of hydroelectric facilities including pumped storage facilities and hydroelectric dams that are located on navigable waters and other waters over which Congress has jurisdiction. The principal standards for granting such a license are that the project is desirable and justified in the public interest and that, in the judgement of the FPC, the project will have a comprehensive, planned approach in using resources and in meeting public needs.

The contents of the FPC license application are set forth in the Commission's Rules and Regulations. The information required includes technical data describing the project, data on economic feasibility, and data on recreational facilities planned in conjunction with the project. In addition, applicants are also required to submit an environmental impact statement covering certain questions raised in NEPA.

U.S. Army Corps of Engineers

The Corps of Engineers under Executive Order 11475, issued by President Nixon on December 23, 1975, was given the authority to establish procedures and issue permits relative to waste discharges into navigable waters and tributaries pursuant to Section 13 of the Rivers and Harbors Act of 1899 (Refuse Act). In conjunction with

Section 10 of that act, the Corps was given authority to approve structures in or crossing navigable waters, including in-flow and outfall pipes, bulk heading, and piers. Since most large generating plants require such structures, Corps permits must be obtained. In fact, the Corps is often the lead federal agency in drafting environmental impact statements for fossil-fueled power plant sites.

Although the authority to issue discharge permits was transferred to the EPA under the NPDES permit program, the Corps is still empowered to enforce the original intent of Section 10 of the Rivers and Harbors Act of 1899, namely to prevent obstruction to navigation. Court decisions, however, have demonstrated that NEPA requirements are applicable to the Corps and that power plants must follow NEPA procedures to obtain such a permit. As a result, the construction permit program requires a complete environmental analysis for fossil-fueled power plants.

Nuclear Regulatory Commission

The Atomic Energy Act of 1954 gave the responsibility of governing the peaceful use of atomic power and regulating the nuclear industry to the Atomic Energy Commission (AEC). This was done to assure that public health and safety were considered. Under the Energy Reorganization Act of 1974, the regulatory functions of the AEC were placed under the Nuclear Regulatory Commission (NRC), and the research activity of the AEC was assigned to the Energy Research and Development Administration (ERDA).

Until the passage of NEPA, nuclear power-plant siting licensing authority was essentially confined to protecting the health and safety of the general public. With the passage of NEPA, and with subsequent Supreme Court Decisions (e.g., the July 23, 1971, decision pertaining to Calvert Cliffs and the 1973 decision regarding Niagara Mohawk), the AEC and its successor, the NRC, were required to consider virtually all direct and indirect environmental effects in siting atomic (nuclear) power plants.

Two license categories exist: (1) commercial and (2) medical therapy, research, and development. Construction of the reactor and associated siting activities may not begin without a construction license from the NCR.

Coastal Zone Management

The Coastal Zone Management Act of 1972, as amended in 1976, is an attempt to balance conservation and development interests in the coastal zone.[24] A federal grant-in-aid program was developed to encourage and support the management of land and water uses of the nation's ocean and Great Lakes coasts. The act vested responsibility for administering the program with the Secretary of Commerce, who has delegated these responsibilities to the Administrator of the National Oceanic and Atmospheric Administration. The Office of Coastal Zone Management (OCZM) oversees the state development of programs and funds such development.[24]

The Coastal Zone Management Act of 1972 contains a planning provision of which several of the requirements directly affect the siting of power plants. Specifically, the act requires:

1. a definition of what shall constitute permissible land uses and water uses, within the coastal zone boundary, that have a direct and significant impact on the coastal waters;
2. broad guidelines on priorities of uses in particular areas;
3. a planning process (a) for energy facilities in the coastal zone or energy facilities that may significantly affect the coastal zone and (b) for anticipating and managing the impacts of such facilities.

It might be noted that most large river basins (e.g., the Ohio River Basin, the Mississippi River Basin, the Colorado River Basin, the Susquehannah River Basin, and the Delaware River Basin) are being studied in a manner similar to the coastal zone management studies.[25, 26] The purpose of these studies is to outline needed land use (including water rights) limits and regulations in each specific river basin drainage area.

Other Agencies

In addition to the five agencies cited above, five other agencies have a special impact in the siting of power plants. The Department of Housing and Urban Development Comprehensive Planning Assistance Program (Program 701) assists in financing planning activities involving local land use impacts. The Federal Aviation Administration assesses hazards to air navigation and is responsible for issuing permits for stacks and towers. The Department of Agriculture issues federal land use permits for power generation and distribution facilities on federal lands, as does the Department of Interior. In addition, the Council on Environmental Quality, an advisory board that focuses on broad federal environmental policy issues, critiques each environmental impact study performed by a federal agency.

While the list of Table 3-3 is not all inclusive, it does present a majority of the federal agencies involved in power plant site approval. Because the number of federal agencies involved in plant siting has grown over the last few years and because increasingly longer delays are being experienced as a result of the involvement by these agencies, the Office of State Programs of the Nuclear Regulatory Commission is attempting to develop new approaches for power plant siting in order to improve both the effectiveness and the efficiency of the process.[27] Some of these ideas are addressed briefly at the end of this chapter.

PRESENT LICENSING PROCEDURES

Hydroelectric Facilities

As noted earlier in this chapter, the FPC has plenary jurisdiction over hydroelectric facilities. Once a project has been licensed by the FPC, either a state or other federal agency cannot prevent construction of the project. In order to qualify for licensing, an Environmental Impact Statement (EIS) must be prepared in accordance with the guidelines noted in Table 3-4.

As a normal procedure, the FPC staff involves as many agencies as possible during the preapplication discussions in order to establish firm lines of communications. Upon receipt of formal application, the EIS, and the exhibits setting forth the recreation and the fish and wildlife facilities proposed for development at the site, they forward copies of this material to the Department of Interior Bureaus of Outdoor Recreation and Sport Fisheries and Wildlife, the Corps of Engineers, and state and county authorities in the area of the proposed project. At this point state and county critiques on plant design are received and evaluated.

Table 3-4 Federal Power Commission Guidelines for Preparation of Environmental Impact Statements.[28]

DESCRIPTION OF PROPOSED ACTION

Purpose
Location
Land requirements
Proposed facilities
- Project works
- Reservoir
- Tailwater features
- Transmission facilities

Construction procedures
Operational and maintenance procedures
Future plans

DESCRIPTION OF THE EXISTING ENVIRONMENT

Land features and uses
- Land uses
- Topography, physiography, and geology
- Soils
- Geological hazards

Species and ecosystems
- Species
- Communities and associations
- Unique and other biotic resources

Socio-economic considerations
Air and water environments
- Climate
- Hydrology and hydrography
- Air, noise, and water quality monitoring

Unique features

ENVIRONMENTAL IMPACT OF THE PROPOSED ACTION

Construction
- Land features and uses
- Species and ecosystems
- Socio-economic considerations
- Air and water environment
- Waste disposal

Table 3-4 Continued

Operation and maintenance
- Land features and uses
- Species and ecosystems
- Socio-economic considerations
- Air and water environment
- Solid wastes
- Use of resources
- Maintenance
- Accidents and catastrophes

Termination and abandonment

MEASURES TO ENHANCE THE ENVIRONMENT OR TO AVOID OR MITIGATE ADVERSE ENVIRONMENTAL EFFECTS

Preventive measures and monitoring
Environmental restoration and enhancement

UNAVOIDABLE ADVERSE ENVIRONMENTAL EFFECTS

Human resources impacted
Uses preempted and unavoidable changes
Loss of environmental quality

RELATIONSHIP BETWEEN LOCAL SHORT-TERM USES OF MAN'S ENVIRONMENT AND THE MAINTENANCE AND ENHANCEMENT OF LONG-TERM PRODUCTIVITY

Short-term uses
Long-term productivity

IRREVERSIBLE AND IRRETRIEVABLE COMMITMENTS OR RESOURCES

Land features and uses
Endangered species and ecosystems
Socio-economic considerations
Resources lost or uses preempted
Finite resources

ALTERNATIVES TO THE PROPOSED ACTION

Objective
Energy alternatives
Sites and locations

Designs, processes, and operations
No action

PERMITS AND COMPLIANCE WITH OTHER REGULATIONS AND CODES

- Permits
  - Authorities consulted
  - Dates of approval
- Compliance with health and safety regulations and codes
  - Authorities consulted
  - Procedures to be followed
- Compliance with other regulations and codes
  - Authorities consulted
  - Procedures to be followed

SOURCE OF INFORMATION

- Public hearings
- Other sources
  - Meetings with governmental and other entities
  - Studies conducted
  - Consultants
  - Bibliography
- Copies of supportive reports

A public hearing is then held before an FPC Hearing Examiner at which time testimony is taken from interested parties. The examiner then issues an opinion recommending that the application be approved or denied. Normally, the FPC will adopt the examiner's opinion.

Fossil Plants

There are no federal regulatory agencies with sole jurisdiction over fossil-fuel plant construction. Case studies on the licensing procedure for these plants indicate that either the Corps of Engineers or the EPA is usually the lead agency in the federal process. A state authority typically takes the lead in the local process.[29] Normally procedures to procure both federal and state permits occur simultaneously. State review, however, usually occurs first.

As an example of a state licensing process, the authors have chosen Kentucky to demonstrate the complexity of the regulatory process relative to fossil plants. Figure 3-1 outlines the review procedure in this state. After a utility has authorized either an expansion or a new facility, enough data must be collected by the utility's design and planning divisions to assess adequately the project impact for all interested governmental agencies. Collected data are used not only for permit applications but also for assessment of the magnitude of the project in terms of resource utilization, economic considerations, and total system improvement.

In Kentucky the utility is required to submit a benefit versus need statement to the Public Service Commission.[30] This statement must be approved prior to any other considerations for permits. (Applications may then be submitted to various municipal, county, state, and federal agencies, as previously listed.)

At the state level in Kentucky, the governor has designated the Department for Natural Resources and Environmental Protection as the agency responsible for processing draft environmental impact statements. Within the department, this function is carried out by an Office of Planning and Research. The goal of the review system is to ensure that environmental impact statements prepared on projects in Kentucky adequately address all areas of local environmental concern.

The Office of Planning and Research performs the clearinghouse function by submitting copies of the draft statements to the state agencies responsible for environmental review. These agencies review and evaluate the statements from their particular perspective. Generally these agencies are given twenty days in which to review a statement and to return it with comments to the Office of Planning and Research. This office then incorporates all comments into an official state review of the draft impact statement and transmits the draft statement to the federal agency that will originate federal review.

Nuclear Facilities[31]

Figure 3-2 presents an overview of the major activities and the major steps involved in the siting process for nuclear facilities. The following sections elaborate on these steps. (Many of these steps also apply to the approval of fossil plants.) The numbers in parentheses relate to the numbered steps in Figure 3-2.

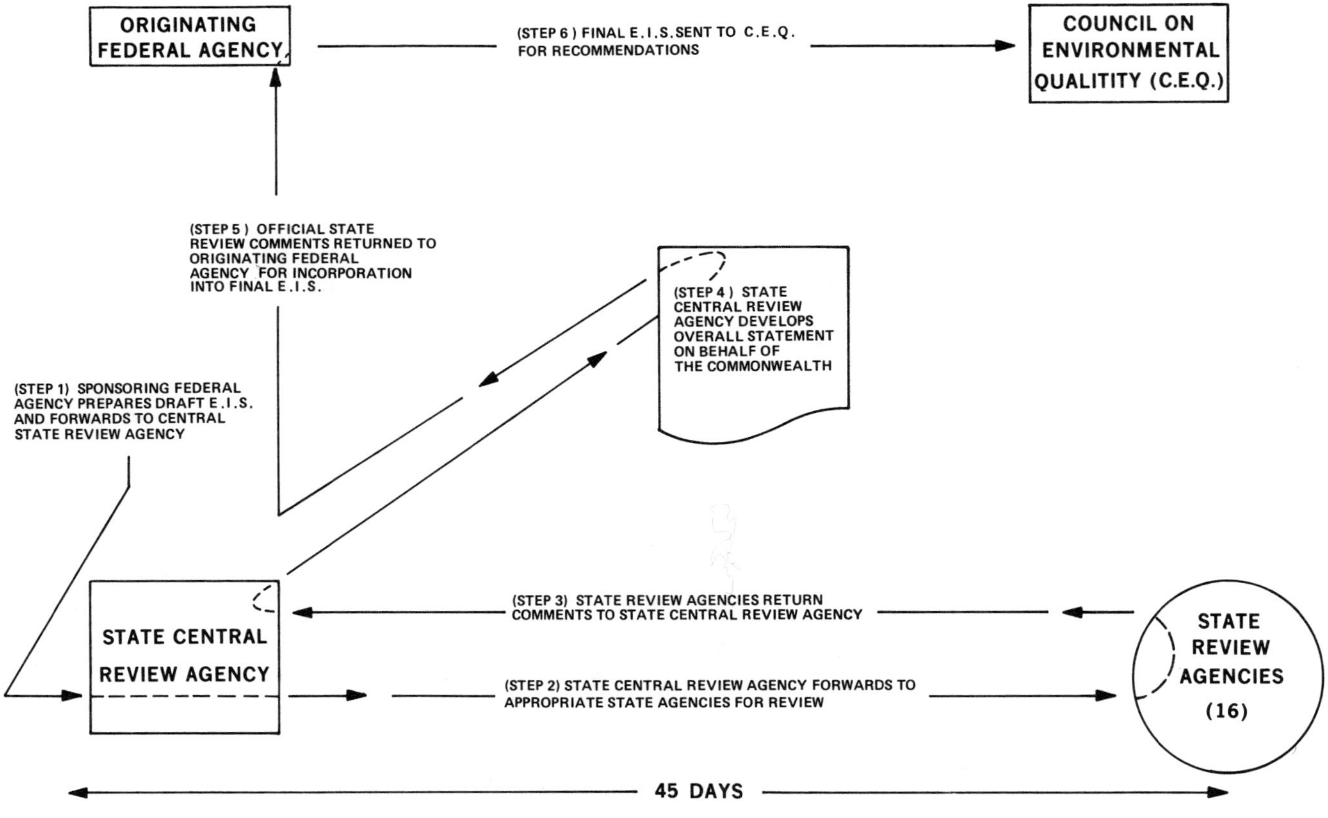

Figure 3-1 Commonwealth of Kentucky environmental statement review procedure.[30]

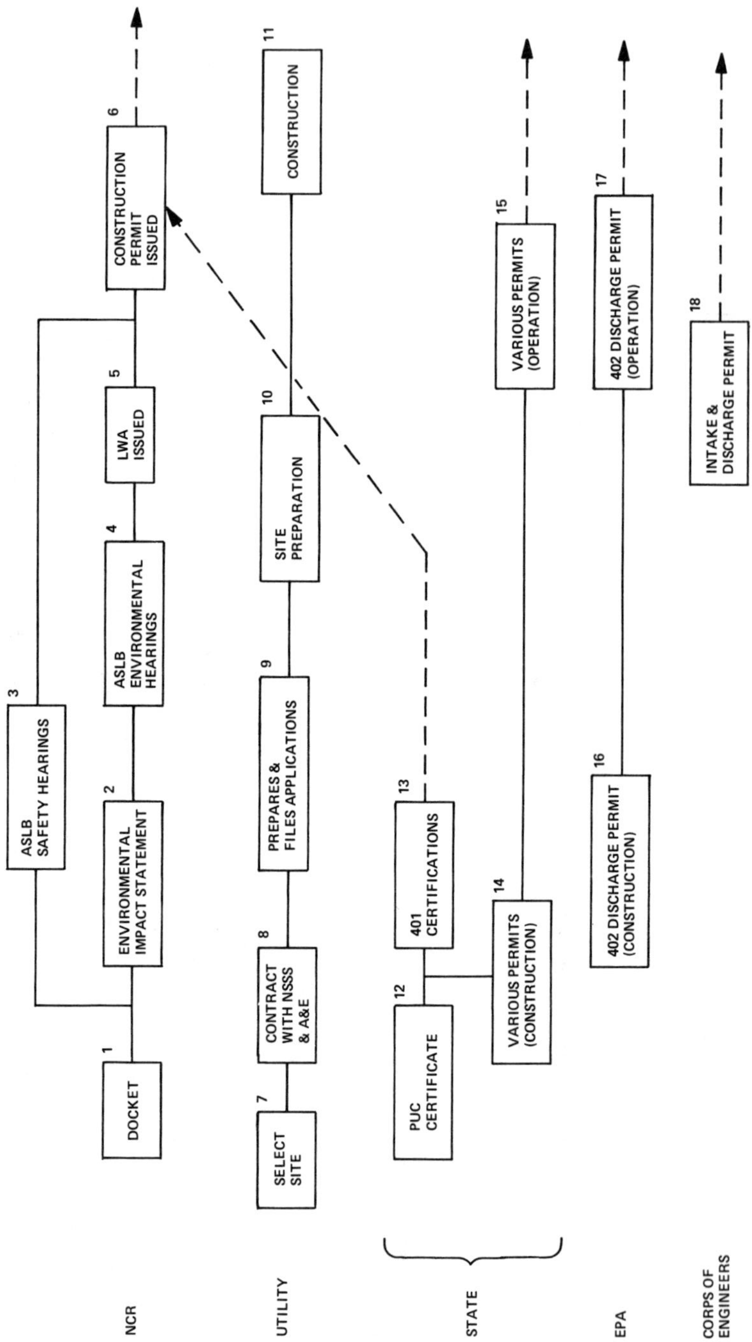

Figure 3-2 Generic nuclear power station regulatory network.[31]

NRC Activities. (1) An application must be filed with the NRC for each facility. This application must include both an environmental and a safety analysis report. (2) The NRC staff supplements the applicant's environmental report with its own findings and issues a draft environmental impact statement. After a comment period, a final Environmental Impact Statement is issued. (3) Safety and (4) environmental evidentiary hearings are then conducted by the Atomic Safety and Licensing Board. (5) In some instances, a Limited Work Authorization (LWA) is granted allowing preparatory site work and allowing limited construction to commence. (6) The final step is the issuance of the Construction Permit (CP). In addition to this process, the NRC also has very extensive requirements for compiling environmental reports. These requirements are detailed in Reference 32.

Utility Activities. (7) Many years before an application is filed, the utility must select the site and secure rights to the property. (8) In addition, the utility must contract (a) with an architect and engineering firm (A & E) for the design and construction of the facility and (b) with a manufacturer for the purchase of nuclear related hardware. (9) The application must then be prepared and filed with the NRC. (10) With the granting of the LWA, site preparation can begin. (11) Major construction, however, must await issuance of the CP.

State Activities. (12) Before construction of a new facility can begin, a state public utility commission must issue a certificate of public convenience and necessity based on the need for such a facility. (13) Also, under Section 401 of the Federal Water Pollution Control Act (FWPCA), the state (or EPA) must certify that any discharge will comply with the requirements of the Act prior to issuance of any permit by the NRC, the Army Corps of Engineers, or EPA. (14 and 15) In addition, state environmental agencies normally must issue construction permits and operation permits for the facility.

EPA Activities. (16 and 17) EPA discharge permits must be issued for both construction and operation in accordance with the National Pollution Discharge Elimination System (NPDES) Program outlined in Section 402 of the FWPCA.

Corps of Engineers. (18) Since most power stations obtain their cooling water from navigable waterways, a Corps of Engineers 404

Dredge and Fill Permit is normally required with respect to intake and discharge structures built in the waterways.

DELAY FACTORS

Generalizations on the sources and the nature of delays in the licensing process of nuclear and fossil plants are difficult to make because of the site specific characteristics of each situation. One cause of delay is that most plants are custom designed. Increased standardization would minimize this delay. Delays in the licensing process can also be attributed to (1) the lack of utility responsiveness to staff or reviewer requests for data, (2) continually changing requirements as to the format for data submissions, (3) understaffed review agencies that contribute to long turn-around times, (4) inordinate amounts of paperwork that must be processed, (5) the complexity of state and federal agencies involved in the siting process, and (6) the recent trend toward intervenor activity.

No conclusive evidence has been developed to establish the specific causes of delay. Continuing study in this area may one day settle upon the specific mechanisms that create delay and lead toward specific recommendations for a more effective and efficient process. A recent staff report from the Office of State Programs of the NPC has been developed, based upon the experience of that agency, that presents suggestions for decreasing delays and improving the licensing and regulatory effectiveness in joint federal/state siting actions.[31] Such evaluations are helpful in creating a public awareness of the deficiencies that must be addressed as viewed from the perspective of a reviewing agency.

REFERENCES

1. "Power Plant Siting in the United States," Southern Interstate Nuclear Board, June 1976.
2. "Power Plant Siting in the United States," Southern Interstate Nuclear Board, September 1974.
3. Beck, R., "Dialogue on Power Demand," The Urban Studies Center, Portland State University, December 1972.

4. "Review of Power Planning in the Pacific Northwest," Power Planning Committee, Pacific Northwest River Basins Commission, Calendar Year 1971, January 1972.

5. Ball, R., et al, "California's Electricity Quandary: 11. Planning for Power Plant Siting," RAND, September 1972.

6. Petruschell, R. and Salter, R., "Electricity Generating Cost Model for Comparison of California Power Plant Siting Alternatives," RAND, January 1973.

7. "California Utilities View Coastal Zoning Differently," *Electrical World*, November 1, 1976.

8. "Power Plant Siting Policy Alternatives for Pennsylvania," Pennsylvania Department of Engineering and Public Affairs, Harrisburg, Pennsylvania, 1975.

9. "Ten-year Site Plans for Electrical Generating Facilities and Associated Transmission Lines," Florida Department of Administration Division of State Planning, Tallahassee, Florida, September 24, 1973.

10. "The Need for Generation in New England: A Strategy to Develop a Modeling Approach for Evaluating Proposed Electrical Facilities," Commonwealth of Massachusetts Energy Facilities Siting Council, Boston, Massachusetts, April 1977.

11. "Local Impacts of Power Plant Siting in Wisconsin," Wisconsin Public Service Commission, Madison, Wisconsin, December 1975.

12. "State Regulatory Activity Involved in Need for Power," U.S. Nuclear Regulatory Commission Office of State Programs, NUREG-0197, April 1977.

13. "Alternative Financing Methods," U.S. Nuclear Regulatory Commission Office of State Programs, NUREG-0204, March 1977.

14. "Success Factor Evaluation Panel," U.S. Nuclear Regulatory Commission Office of State Programs, NUREG-0196, March 1977.

15. "Nuclear Power Plant Licensing: A New England Perspective," U.S. Nuclear Regulatory Commission Office of State Programs, NUREG-0202, March 1977.

16. Fellman, D., et al, "CAPS: Comparative Analysis of Power Siting in the United States," Ohio Power Siting Commission, Columbus, Ohio, June 1, 1977.

17. Evans, V., "Recent State Laws Regulating Power Facility Siting," *Nuclear Technology*, Vol. 31, December 1976.

18. "Laws and Procedures of Power Plant Siting in New England and New York (Revised)," New England River Basins Commission, March 1972.

19. "Energy Facility Siting: A Comprehensive Program for Pennsylvania," Pennsylvania Joint State Government Commission, Harrisburg, Pennsylvania, April 1977.

20. *Proceedings of the 1975 Federal-State Conference on Power Plant Siting,* U.S. Nuclear Regulatory Commission Office of International and State Programs, July 1975.

21. *Proceedings of the Second State-Federal Power Plant Siting Conference,* U.S. Nuclear Regulatory Commission and the National Governors' Conference Energy Project, June 16-18, 1976.

22. Lippek, H., "Power and the Environment: A Statutory Approach," *Washington Law Review,* Vol. 47, pp. 35-72, 1971.

23. Rogers, W., "Siting Power Plants in Washington State," *Washington Law Review,* Vol. 47, pp. 9-33, 1971.

24. "Environmental Planning and the Siting of Nuclear Facilities: The Integration of Water, Air, Coastal, and Comprehensive Planning into the Nuclear Siting Process," U.S. Nuclear Regulatory Commission Office of State Programs, NUREG-0199, February 1977.

25. Cole, S., "Ohio River Basin Energy Study," *ORBES,* October/November 1976.

26. Hauck, F., "Unneeded Generating Plants in Ohio River Valley," Unpublished Report Prepared by Author in Louisville, Kentucky, November 1976.

27. "Detailed Study Plan Efficiency in Federal/State Siting Actions," U.S. Nuclear Regulatory Commission Office of State Programs, NUREG-0128, October 1976.

28. "Resource and Land Investigations (RALI) Program: Methodologies for Environmental Analysis, Vol. 2, Power Plant Siting," MITRE Corporation, MTR 6740, August 1974.

29. Philbin, T., "Impacts of NEPA and Other Court Actions on Fossil Fuel Option," Ecological Analysts, Inc., January 11, 1977.

30. "Energy for Industry in Kentucky," Commonwealth of Kentucky Department of Commerce Research and Planning Division, Frankfort, Kentucky, 1970.

31. "Improving Regulatory Effectiveness in Federal/State Siting Actions," U.S. Nuclear Regulatory Commission Office of State Programs, NUREG-0195, May 1977.

32. "Regulatory Guide 4.2, Revision 1: Preparation of Environmental Reports for Nuclear Power Stations," U.S. Nuclear Regulatory Commission, January 1975.

4

Siting scenarios

INTRODUCTION

Siting Evolution

From its beginnings in the 1800s, the electric power industry's methodologies and rationale of central generating station planning and operation have closely paralleled the advance of our developing social organization. At the beginning of the technological expansion, the primary objective of siting a plant was to find a location that provided power at the lowest possible cost. In order to reduce the complexity and design of the transmission system and to appear physically more visible and dependable to their largest customers, generating facilities were usually located in, or near, the industrial sections of towns and cities. This siting usually resulted in access to industrial water supplies and easier fuel transportation (by barge or railroad). The opportunities afforded by this proximity to the utility's load center, combined with the small generating stations of municipal vintage, provided the main rationale in the early sitings.

The growth of our communities preempted land acquisition for future utility expansion as the loads increased. Zoning code restrictions often excluded power plants from prime industrial sites. The continued need for a dependable power supply forced the electrical utilities to move to the suburbs. Consequently, transmission line investments and the subsequent improvement of distribution systems resulted in higher capital investments. However, increased efficiency of new plants and lower taxes offset these costs.

As the expansion of cities continued and demand for land and water resources became more competitive, utilities were forced farther and farther from the load centers. Economic trade-offs were required to offset the greatly increased transmission and delivery costs. Higher transmission voltages were utilized. Increased equipment efficiencies and "economies of scale" were used to generate larger quantities of power at decreased incremental costs. Fuel and transportation costs were minimized by locating power plants near, or easily accessible to, fuel sources. The siting methodology evolved to defining the most economical area in which to construct a plant, reconnaissance of this area to locate a suitable topography, and acquisitioning of as much land as possible at the lowest cost possible.[1, 2]

As prime power plant sites were used, the search for future sites became more complex and difficult. Once-clear economic trade-offs and technical parameters became clouded by social pressures and industrial competition. With the environmental movement of the sixties, siting based strictly on economic, technical, and socio-political alternatives was not sufficient. In addition to these considerations, environmental compatability was also required.

Today the system planner must evaluate not only the impact of the proposed facility upon the environment, but often he must change the plant design in order to mitigate undesirable impacts. In addition, the utility's plans are often placed under detailed public scrutiny. Engineering alternatives and designs, once the exclusive domain of the system planner, are now questioned by environmentalists, politicians, and the public. In many cases individuals reviewing the proposed facility are unable to grasp either the basis for the technical designs or the rationale behind siting and design decisions.

The literature abounds with reports on general power plant siting methodologies, criteria, parameters, impacts, and procedures.[3–11] Annotated bibliographies exist that list articles and studies on each aspect of the siting process. Comprehensive reviews and specific case studies are available on historical plant siting situations. While the scope of this literature is quite diverse, there is a general consensus within the literature that three primary factors govern all power siting procedures today. These factors are commonly known as the "three e's" of power plant siting: engineering, economics, and environment.

The Siting Interaction

Engineering design must be tempered by economics to produce a facility that is compatible with the environment. This interaction is basic to each power plant siting action. Since the overall siting process is dynamic, these three influences are constantly interacting. The reader should keep in mind that it is difficult to completely separate these three components.

The purpose of this chapter is to present and discuss many of the general factors and options of the present power plant siting process. While this treatment is in no way intended to be a detailed technical explanation, an attempt is made to focus on several pertinent design issues and their associate design options. Since economic considerations dictate, to a large degree, the final decisions, these factors will be discussed where appropriate. The main thrust of this chapter, however, will be to view the siting process primarily from the perspective of engineering, the basic design criteria for a power plant.

Nature's endowments of fuel, water, and land are all interwoven into the fabric of the power plant siting process. These three factors best illuminate the engineering perspective. Fuel quality and availability dictate the type of plant to be built. Availability of water mandates the plant's cooling system design, and land features characterize plant location. Each of these general parameters will now be discussed.

FUEL

Throughout the siting process utility engineers are forced to make design decisions on the proposed facility. All initial decisions relate to the choice of fuel supply. As discussed in Chapter 6, many options are available to the system planner. Hydroelectric, pumped storage, coal-, gas-, oil-fired, and nuclear plants are each utilized to meet the specific needs of the operating system.

The art of planning and expanding a utility system is known as capacity expansion optimization. Many variables are considered in this process. Often computerized models are used to solve involved theoretical optimization calculations. In these analyses, system capacity and fuel type requirements are balanced against economics. In addition, the mechanics of the fuel choice must also be examined from a physical siting viewpoint. Power siting requirements and trade-offs must by analyzed and compared for each alternative fuel.

This section will feature a brief discussion of the major fuel alternatives and their siting constraints from a physical point of view. Discussion will be restricted to fossil-fired, nuclear-fired, and hydroelectric/pumped storage generation schemes. Hopefully, this discussion will provide the reader with valuable insight to the entire siting process.

Fossil-fired Limitations

Historically, coal was the mainstay of utility boilers. Over the past ten years, however, fossil-fired plants shifted from coal to oil and natural gas. This shift was due primarily to both the environmental restrictions placed on the stack gas emissions from the boilers and the available, low cost alternate fuels. Today, with federal emphasis on the growing shortage of domestic oil and natural gas, fossil-fired plants have shifted back to coal.

The use of coal as a fuel supply presents two specific constraints on the site to be chosen. First, the site must contain an area sufficient to store the coal prior to its actual use. Second, since the burning of the coal leaves a slag or ash behind, this solid waste together with the particulate removal accumulations, must also be stored onsite until disposal. If disposed at the site, 250 to 400 additional acres of land is required for settling ponds and basins. These constraints are in addition to the land required for actual plant construction, fuel handling facilities, cooling system facilities, and environmental control facilities, which require a minimum of 600 acres.[12]

Health science concerns about aggregate air and water emissions have restricted coal-fired plants in highly developed areas. Although this type of siting problem can be overcome, the engineering cost usually proves uneconomical. Currently, questions are being raised about the aggregate effect of the air and water effluent discharges from plants in close proximity; thus, there must be limits to the proximity of a plant to industrial complexes.

Finally, the availability, accessibility, and type of coal are all important features in site selection. Coal barge transportation, unit-train railroad shipment, and coal-slurry pipeline commitments are all viable alternatives that must be reviewed. With increased efficiencies of transmission, "mine-mouth" plants having direct access to the fuel source through conveyor systems or truck shipments are possible alternatives. The type of coal available also influences site selection. Since sulfur content will dictate the type of air emission clean-up system, availability of low sulfur coal can be an improvement feature.

Hydroelectric/Pumped Storage Limitations

Today, utility dependence on hydroelectric generation is diminishing. In the northwest and in parts of the south hydroelectric generation is still a viable alternative as a result of federal government programs instituted years ago (e.g., the Bonnieville Power Administration (BPA) of the northwest and the Tennessee Valley Authority (TVA) in the south). While some additional hydroelectric capacity is projected in certain regions of the country, there are relatively few remaining undeveloped sites that can be used effectively.

Hydroelectric generation facilities are primarily located on rivers. Since dams and spillways must be constructed to acquire the water pressure and flow necessary to generate electricity, favorable topographical conditions must exist to minimize construction expenses. Natural construction points on a river, which are narrow and acceptable for construction, are difficult to find. Often land use constraints eliminate many possible sites. The environmental impact the construction of such facilities creates is also a concern that often prevents the construction of major projects.

Water power, however, will continue to be used. In the future this energy source will take the form of pumped storage facilities. The concept behind pumped storage is to utilize off-peak energy (energy generated during the time of day of least electrical demand) to pump water into a reservoir at an elevation higher than normal. Then, during periods of peak demand, the water can be used to generate hydroelectric energy and meet customer demands. Though these facilities are less than 100% efficient, they are available for peaking capacity and to level system generation requirements.

Although hydroelectric plants are restricted to conditions that favor locations directly in rivers (adequate headwaters and flow), pumped storage facilities can be located off waterways. Again, appropriate physical conditions of the surrounding terrain are necessary for constructing a reservoir. Precautions against environmental damage must be made. However, land requirements are not as extensive as in the case of fossil plants.

Nuclear Plant Limitations

Commercial electric power generation by nuclear fission began in 1957 with a prototype 90 megawatts pressurized water reactor at Shippingport, Pennsylvania—an AEC demonstration plant based on nuclear submarine technology.[13] In the ensuing years, nuclear reactors have developed in size, complexity, cost, construction time, and licensing requirements. Today, nuclear plants are a viable system candidate.

Since nuclear plants do not require daily fuel loading or supply, there is no need to maintain a fuel storage backup. The plant, along with accompanying storage facilities for spent fuel rods and auxiliary plant facilities, requires only 80 to 100 acres, roughly a factor of ten less than a conventional fossil-fired plant. However, due to licensing requirements and safety constraints, an exclusion area of from one to five miles, depending upon population density, must be maintained around plant sites.[14] Also, minimum distances from population centers are required in order to provide additional safety clearances.

The cooling system for nuclear plants will be larger and more expensive than for fossil-fuel facilities due to the lower efficiencies of the nuclear plants. "First-generation" nuclear plants are about 33% efficient. Future reactors, using advanced design, may increase this efficiency. However, even then the nuclear cycle will probably operate at a higher output and thus require additional cooling facilities. If cooling towers or lakes are not viable, alternative systems such as cooling ponds and canals, which can result in the extensive conversion of existing land use on the order of thousands of acres, may be required. Since nuclear plants have a potential for extensive thermal pollution, detailed consideration is given to the cooling system in the siting activity.

One final factor influencing the choice of the nuclear option is accessibility. Delivery of heavy equipment and large shop-fabricated and assembled reactor vessels presents transportation problems. The very size of this equipment often necessitates shipment by water. Additionally, safe and reliable transportation of expended fuel rods must be guaranteed. While transportation of the nuclear fuel to the plant entails no unusual problems, once the fuel is used, radioactivity is inherent in its structure. Transportation of the expended fuel to reprocessing or storage facilities presents special safety hazards. For these reasons transportation facilities play an important role in nuclear plant siting.

Although there are many more variables considered than those discussed in the sections above, the variables presented are the primary site specific constraints on fuel alternatives. One common aspect to any siting study is the requirement for a reliable water supply. This requirement, the second major building block of a plant site, will now be discussed.

WATER

Many site specific problems can be overcome by engineering and design. However, the necessity for an adequate supply of water eliminates many otherwise acceptable power plant sites. Electricity is created by generating steam at a relatively high pressure in a boiler. When the steam passes through a turbine, work is performed. Energy is removed from the steam thereby reducing its pressure and temperature. On leaving the turbine the steam is condensed to water by the cooling water system.

The condensation of water enables marked efficiency increases in the generation cycle. A basic law of thermodynamics relates the efficiency obtainable by any heat engine to the temperature limits of the cycle. Thus, the cycle efficiency is increased by raising initial steam temperature and by lowering the exhaust temperature. The function of the cooling water system then is to reduce the turbine exhaust temperature so as to obtain high cycle efficiencies and conserve fuel.[15] Figure 4-1 presents the essential cooling system components present in a steam-electric power generation system.

Hydrological investigations concerning power plant sites involve two considerations, namely (1) the availability of water for steam generation and for condenser cooling purposes and (2) the actual design required for the cooling water system itself. Initial investigation is usually limited to determining the sources of water that exist and the quantities of water available from these sources.

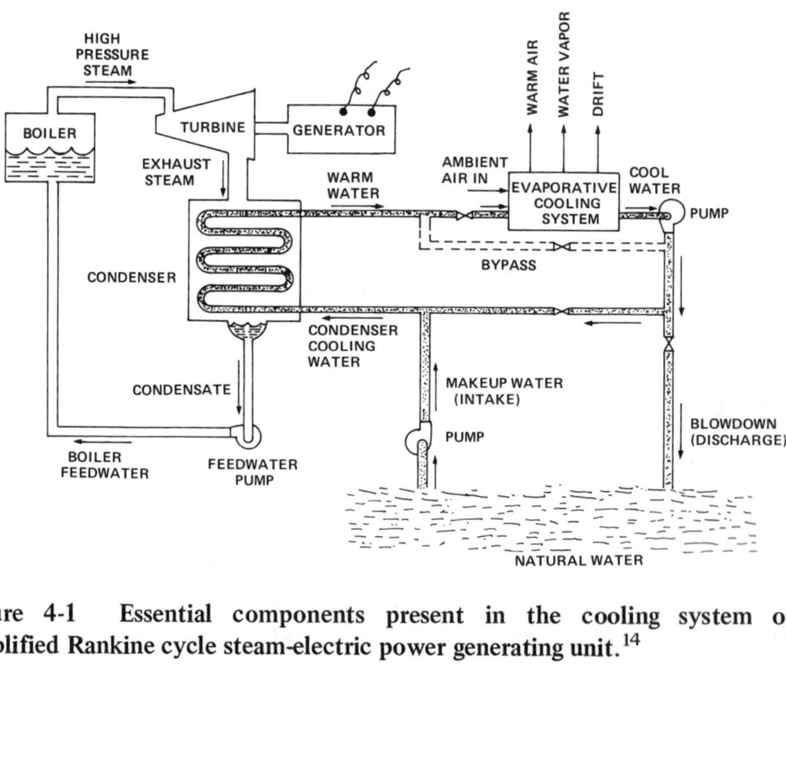

Figure 4-1 Essential components present in the cooling system of a simplified Rankine cycle steam-electric power generating unit.[14]

Availability

Rivers, lakes, coastal regions, reservoirs, and groundwater are all potential water sources for a power plant. Each source possesses particular advantages and disadvantages. For instance, the use of groundwater, while offering easy access in practically any situation, is limited by the withdrawal rate and the amount available in subsurface aquifers. Although a viable alternative for medium sized plants, groundwater supplies cannot approach the needs of large generating complexes.

When considering sites along rivers and estuaries, detailed records are required to determine the extent of river resources. Adequate water supply must be assured during periods of low flow. Aggregate use of water from upstream and downstream consumption must also be predicted, especially when there is common use by industries and communities for irrigation, transportation, and sewage disposal operations. While these waterways are considered to be renewable resources with the ability to assimilate and purify discharges, this very capability breaks down in the face of excessive chemical and solid waste discharge. Therefore, purification of inlet and discharged water is often necessary in the design of a system.[16] Finally, thermal discharges to the rivers and estuaries, although limited by cooling water system design, is a constraint upon the utilization of these water bodies. Care must be taken to predict and measure the water temperature and its variations at specific site locations in order to properly design an efficient and acceptable system.

Lakes and reservoirs present different types of siting problems. Traditionally, this type of freshwater body, whether natural or man-made, has been used for drinking water and waste disposal processes. In addition, the use of lakes and reservoirs for recreation, industry, transportation, and commercial fishing can add constraints on plant siting. The maintenance of the quality and biological balance of these bodies of water, however, is the limiting factor in the siting consideration. For example, while a lake shore site can offer the possibility for an open system for cooling purposes (a once-through cooling system using the extremely large heat sink of the water body for economical cooling, thereby minimizing land requirements), the total volume of water usage could be sufficiently high to destroy these balances.

Although availability of water from lakes and reservoirs possesses certain limitations, the heat sink of the water body has an economic attraction. Consider then an even larger heat sink available for power siting purposes: the ocean. Siting in coastal regions negates many of the common use problems encountered with lakes and reservoirs. However, the thermal effects of the discharged effluent must be more carefully studied, for they can produce more adverse effects. In addition, design of the cooling system is hampered by lack of freshwater supply.

Meteorological Effects

The need for an adequate water supply often necessitates year-round monitoring efforts to predict seasonal meteorological variations. One seasonal variation essential in maintaining a reliable water supply, yet difficult to predict, is rainfall. On rivers and estuaries, periods of heavy rainfall raise the spectre of plant site flooding. Thus careful thought must be given before using flood plain and delta sites. On lake shores and reservoirs high winds can cause wave action and damage to structures. Additionally, current variations can affect discharge operations. With coastal siting, tidal effects and wave patterns must be known in order to safely design a reliable system. Since meteorological effects play an important part in determining the siting of plants along these water bodies, they must be considered throughout the siting process.

Cooling System Alternatives

Once the availability and quantity of water has been assessed, a review is performed relative to the design of a cooling water system that will combine the necessary plant parameters with the available resources. Although this assessment and review is normally performed concurrently under a broad general screening program, this presentation will consider the various system alternatives relatively independent of availability characteristics. The various alternatives that exist are (1) once-through cooling, (2) cooling towers, and (3) cooling lakes, ponds, and canals. Combinations of these systems may be used. For example, a cooling tower can be augmented with a lake for periods of low flow. We will, however, restrict this discussion to the basic alternatives just noted.

Once-through Cooling. There are four basic means by which heat is transferred from water to air: conduction or convection, evaporation, radiation, and advection. Once-through cooling systems utilize the natural ability of the receiving water body to dissipate the heat by these natural processes. A once-through system describes an arrangement which draws water from a natural water body, utilizes the water throughout the condenser system, and then discharges the water back into the original source. Thus, the water passes through the system once and then is discharged.[17]

There are two parameters used to design any cooling water system. The first is the withdrawal rate of the system – the amount and rate of water drawn into the system through normal operations. The rate of a once-through system is much higher than the rate of a closed system. The second parameter is the evaporation rate of the system. With a once-through system, the evaporation rate is obviously the rate of the receiving body. For a closed system the proportion of the total heat that is transferred from the cooling water by evaporation varies according to the system design. While a once-through cooling system requires a large body of water to achieve proper heat dissipation, the loss of water through the evaporative process is usually less than in a closed system.[12]

Because of the large withdrawal rates in once-through systems, the design of the intake system is important. Screens must be used to prevent the blockage of condenser tubes or pumps by solid materials in the water. In addition, as warm temperatures enhance the growth of marine organisms, chemicals must be added to maintain proper heat transfer surfaces throughout the system.

Discharge techniques must also be carefully designed since the discharges are scrutinized by regulatory bodies. The addition of materials, chemicals, and heat to the receiving body must be minimized so as to prevent detrimental effects on the biology and quality of the water.[15, 16] The rate and velocity of the discharge flow must also be controlled to prevent adverse effects on organisms close to the outlet. The primary concern, in the design, however, is the thermal plume. Once-through systems are normally restricted to large bodies of water because of this plume.

Cooling Towers. A cooling tower is an apparatus used to enhance the dissipation of heat from the cooling cycle medium. The rate of heat transfer in a cooling tower, for both mechanisms of evaporation and convection, increases with corresponding increases of (1) air to water interface surface, (2) relative velocity, (3) contact time, and (4) temperature differential. The first three parameters can be affected by cooling tower design (type of fill, type of fans or chimneys, tower size, etc.). The variation of the temperature differential is a function of atmospheric conditions.[9]

Cooling towers are closed systems. Since a closed-cycle system is able to reuse most of its water, it needs to replace only that water lost to the atmosphere by evaporation. The water source used to replace this lost water is called the "make-up" water system. The water system, the collector system, and the recirculating system are all concerns in the design of a cooling tower system.

There are many types of cooling towers available. In the "wet" type of tower the water to be cooled comes into close contact with the cooling air and transfers about 75% of the total heat by evaporation. This transfer occurs in the fill, two types of which are used today. The splash bar type uses a series of horizontally staggered bars to break the water into droplets as it falls by gravity through the fill and cooling air. The film type uses a series of vertical plates to form the falling water into thin sheets between which the cooling air can pass.[17]

The water in a cooling tower can be drawn up through the tower by natural or mechanical means. It is then drawn back down through the fill by gravity. The relative flow of the air and the falling water may be described by two different terms. Horizontal air flow, crosswise to the vertical falling water, is known as "cross flow." When air flow is vertically upward, counter to the downward falling water, it is called a "counter flow." In "natural draft" towers the air is induced to flow upward by means of a large chimney that maintains a density differential between the air inside and the air outside of the chimney. Air movement through a "mechanical draft" unit is accomplished by fans. (See Figure 4-2.) These fans may be either the induced draft type (ID) that pulls air through the tower or the forced draft type (FD) that pushes air through the tower. Combination mechanical and natural draft systems incorporate both a chimney and fans.[12]

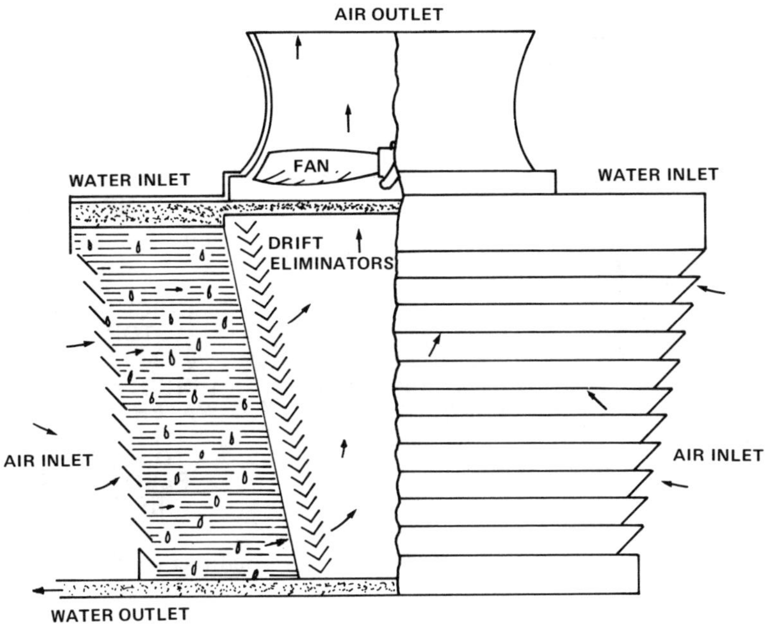

Figure 4-2 Mechanical-draft cross-flow evaporative cooling tower.[17]

The dry cooling tower arrangement, while reducing the amount of plume from the tower, requires additional power to operate extra pumps and fans. The circulating water never comes into direct contact with the air but is cooled by convection from pipes within the chimney. (Figure 4-2 presents a schematic of an indirect dry cooling tower.) In areas where water supplies are available for wet towers, the capital investment for dry towers is prohibitive. However, a lack of water would necessitate this use of a dry tower.

The limitations upon design considerations of these towers range from aesthetics and economics to impact from both the plume and discharge into the receiving body. A tower may be designed to limit the thermal effect of its plume and discharged water. However, there is an economic trade-off in capital investment and power requirements. A balance is achieved by considering local conditions and allowable plant parameters.

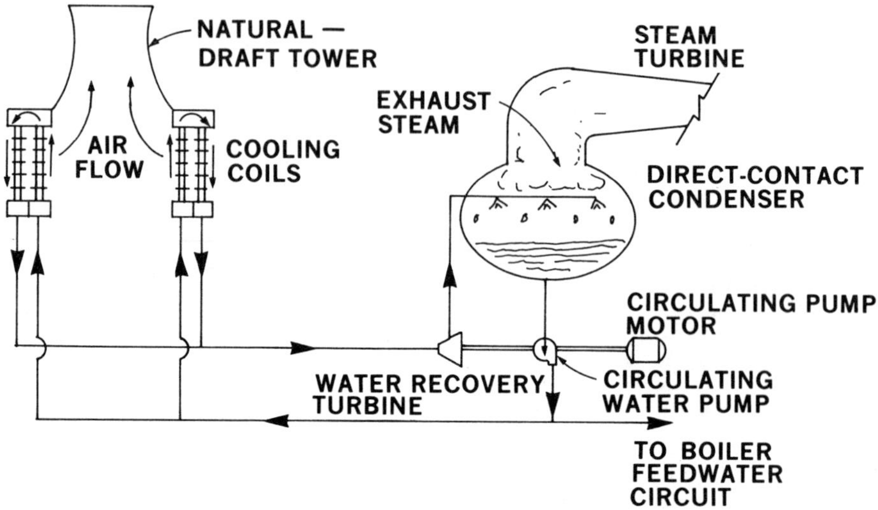

Figure 4-3 Dry cooling system design for indirect cooling.[16]

Cooling Lakes, Ponds, and Canals. Where natural lakes are not available and river flows are inadequate, man-made cooling ponds or canals can be constructed (see Figure 4-4). The design characteristics of these artificial lakes include the density of the water, the specific heat of the water, the heat transfer coefficient of the receiving body, and the equilibrium temperature of the receiving body. There are three general types of cooling ponds, namely (1) completely mixed lakes that have the same temperature throughout, (2) flow through lakes where the temperature decreases away from the warm water discharge point (a canal would operate in this manner), and (3) internally circulating lakes where warm water flows on the surface while cool water flows to the intake underneath. Depending upon land requirements and thermal and capital constraints, any one of these lakes may be utilized.[17]

Spray canals and ponds are similar in construction. (See Figures 4-5 and 4-6.) They are both usually designed to have perforated pipes or some mechanical means, such as floating aerators, to form water droplets. Thus, air–water contact is enhanced and evaporative heat transfer occurs by spraying the droplets into the air. Spray pond design will vary depending upon climate conditions, normal wind speed, and land availability.

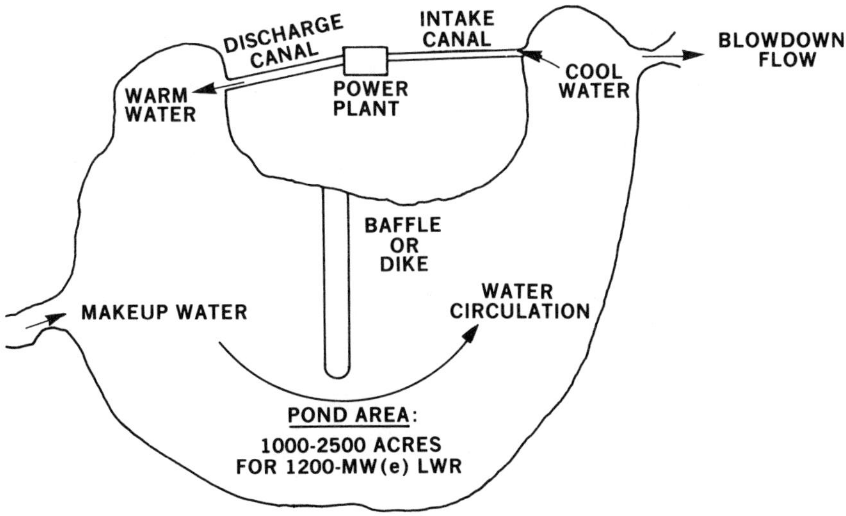

Figure 4-4 Cooling pond system.[14]

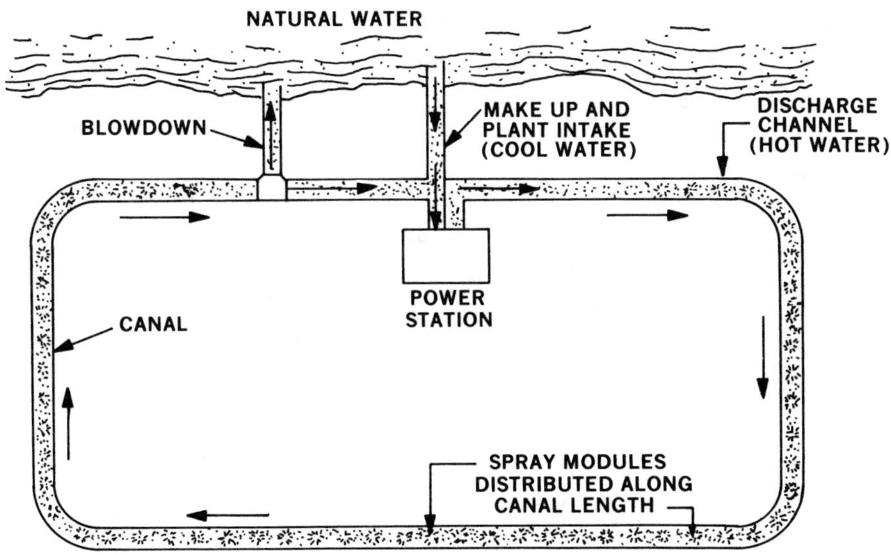

Figure 4-5 Spray canal schematic.[16]

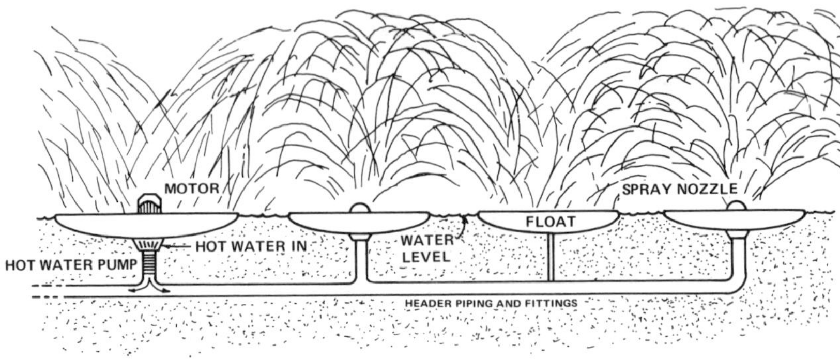

Figure 4-6 Illustrative spray cooling system.[16]

Use of cooling ponds and lakes requires a knowledge about the total plant design. Input and discharge water temperatures must be calculated in order to assess the efficiency of the system and its impact on the receiving body. Since both the temperature of a lake and the load on the power plant will vary throughout the day, thermal characteristics will continually change. Because these changing characteristics can have adverse effects on a closed-cycle system, the maximum level to which water temperature can rise must be projected as closely as possible and considered in the plant design. Often these studies may be computerized allowing interaction of a large number of variables.

Other factors can play an important part in the design of each of these systems. An artificial lake may have a large surface area and shallow depth, thus saving construction costs. Or, if available surface area is limited, a deeper lake can be constructed. The length of canals may be dependent upon nonland use factors such as ambient conditions and the type of plant being constructed.

Blowdown

In addition to the water required for actual cooling, water is also required to cope with a phenomenon unique to all closed cycle cooling systems, namely the tendency for dissolved solids to concentrate. As water evaporates, solids are left behind. New water introduces more solids thereby increasing concentrations. Thus, some of the solids must eventually be discharged. The discharged water containing the concentrated solids is called "blowdown." The design for the blowdown operation must consider the level of solid concentrations anticipated and the volumes of water required for operation.

LAND

Land, the third major engineering consideration in power plant siting, will now be considered. After the choice of fuel has been made and after water availability is established and cooling system economies identified, a precise location for the power plant can be determined. The term "land," as used here, includes all engineering consideration dealing with the physical characteristics of a site. The more prominent parameters are geology, topography/geography, demography, and accessibility.

Geology

The geological investigations for a power plant site start with a study of the type of bedrock and basement rock in the plant environs. Data from past surveys, and new core drillings are used. Normally, the geology will not vary over a large region. However, as each specific site is considered, numerous core samples and structural tests are undertaken to establish local conditions. These geological investigations are necessary to determine the capability of the ground to support the heavy structures and foundations of the power complex.

In the investigation process any mineral deposits or geological hazards (such as abandoned mines or wells) are identified. This information is important to foundation design since hollows left by extraction of these deposits often result in subsidence of the surface over a period of time. Since subsidence would be detrimental to the plant, these regions must be avoided or fortified by grouting and filling the subsurface fractures.

Another aspect of the geological consideration is the tectonics or seismology of the area. The possibility of earthquake conditions, including age and distribution of faults, is investigated. This study usually includes a review of the literature and historical records of past activity in or near the area. Also, an attempt is made to identify any site specific fault conditions by field tests. These considerations are most important to the siting of nuclear plants because of strict tolerances set forth by the NRC.[11]

Topography/Geography

In this area of investigation an attempt must be made to identify the actual physiographic characteristics of the proposed site. Given the degree of engineering expertise available today, it is unlikely that there are many unsuitable places. However, economics takes a prominent role here.

A preferable power plant site, from a topographical viewpoint, would possess relatively mild terrain conditions and adequate room for future expansion. One would not attempt to construct a plant on rough terrain that would require extensive activity to conform the topography to desired grades. The economic cost would be unjustified. Another topographical consideration would be the presence of natural valleys for impoundment construction. Specific topographical requirements will be a "parameter interrelated function" of the type of fuel and cooling system selected for the plant.

Demography

When considering the proposed location of a plant from a demographical viewpoint, the designer is concerned with population densities, existing land use, proximity to loads, and available manpower. Nuclear power plants are an exception since they are often situated in secluded areas. On the other hand, since the construction of a power plant can require from six to ten years and involve several thousand people during the maximum construction period, the availability of manpower for employment is important. The demographic study must then consider construction impacts on the local economic structure, availability of public services, and the availability of specific construction trades. These elements are treated in more detail in the next chapter; however, their combined interest is evaluated here from another perspective, the availability of land in the increments required.

Numerous studies have been performed in each region of the country to determine proper land use patterns. These studies have been especially prolific for water basins. Some of these studies have been performed in a depth sufficient to enumerate and predict the consequences of siting a power plant in the region.

Competition for land along the shores of lakes and estuaries is an integral concern of the power siting process and is becoming increasingly important as the land around these water bodies is developed. The regulation of land development in water basins by planning commissions is also a concern. As communities, counties, or states develop new guidelines for population growth, these guidelines must be considered in order to ensure that plant location and plant impacts do not violate these plans. As land is zoned for agricultural, residential, or industrial uses, siting plans must be altered accordingly.

Availability of land is an important aspect of the process. The important features here are not only physical availability but also the economic availability. The viability of purchasing the needed land rises with a decrease in the number of landowners. The very nature of the proposed facility precludes competing land development. Therefore, the land must be committed to one purpose, the generation of power.

The proximity of the plant to load centers is also an important parameter. Forecasts of energy and demographic patterns are used to locate future load centers. Although often constrained by other limitations, a plant site close to a load center is preferred to one remote from the load center. This philosophy reduces the cost of transmission lines and associated rights-of-way for transmission facilities and improves system economics.

Accessibility

As mentioned earlier, a plant site needs to be accessible for fuel shipments, in the case of fossil-fired plants, and for certain prefabricated components, in the case of nuclear plants. Sufficient highway, rail, and river facilities must also be available for transportation of construction workers and supplies to the site. In addition, public services such as utilities, fire protection, and emergency medical services must also be accessible to the site. (These services and the plant's impact upon their availability are discussed in the next chapter.)

Another aspect of accessibility is the site's ability to sustain exit corridors for transmission facilities. Without these escape corridors generated power cannot be economically transmitted to the load centers where the power is needed. Many good plant sites are totally surrounded by sensitive land-use classifications and thus have no visible means of exit from the plant. By including this parameter early in the siting process, situations of this type may be avoided.

METEOROLOGY AND NUCLEAR CONCERNS

The aspect of meteorology is evident throughout the power siting process. Special mention of this topic is made here in order to point out that while the system planner cannot control the weather, specific meteorological factors can be included in site selection.

The planning process is usually based upon a "worst case" philosophy. In other words, an attempt is made to predict the worst possible case that might arise and to design the plant based upon this phenomenon occurring in a given situation. Wave action, storms, flooding, and high winds affect plant parameters and system design. Worst case conditions for each must be anticipated in the design. In the case of a nuclear plant, the reactor building and associated facilities must be able to withstand not only normal natural forces but also abnormal forces such as tornadoes, hurricanes, and earthquakes.

Another nuclear facility concern relates to radioactivity levels. Waste disposal systems for both high-level and low-level nuclear products must be incorporated in the plant site design.[18, 19] Effluents result from these gaseous, liquid, and solid radwaste. Because of the radioactive nature of these effluents, special monitoring devices must be incorporated in the plant design. These systems are very detailed and must all be approved by the licensing authority.

Safety analysis of the nuclear plant also plays a part in the siting process.[20] Because of the catastrophic events that could result from a nuclear core meltdown or major system failure, special safety engineering features are required by the NRC. The NRC espouses a policy of "defense in depth." Backup systems to the immediate area and multiple pathways of safe core shutdown and containment are involved in this process. Only through proper design, monitoring, and redundancy can safe systems be developed. A newly evolving safety concern relates to the aspect of plant and system security. Future nuclear plant designs will require increased security from operator negligence and external sabotage.

Finally, several alternatives to nuclear power plant siting have been offered. Two of these alternatives are underground and offshore plant siting.[21–23] Some of the advantages of the underground alternative include greater safety and shielding in the event of a nuclear release, reduced surface area requirements, and improved aesthetics of the plant environs. While offshore plant siting has been a viable alternative, coastal zoning authorities are skeptical as to the supposed advantages of such a facility (e.g., separation, reduced cooling costs, and safety) and afraid of the consequences of increased thermal and radioactive effluents to the ocean. Nevertheless, four floating nuclear power plants have been ordered by the Public Service Electric and Gas Company of New Jersey. (The first unit is scheduled for delivery in 1984.)[24]

CONCLUSION

This chapter has presented some of the basic system design alternatives available to the power siting engineer. A purely technical presentation was avoided because of space limitations. Readers desiring more technical depth should consult the references at the end of the chapter.

The technical perspective for power plant siting can be obtained by considering the three major siting influences: fuel, water, and land. Each influence contains numerous alternatives with associated economic effects. The engineering design team must consider each alternative, and its economic effect, before selecting a specific site design.

This chapter emphasized engineering concerns, as influenced by economics, without considering the environmental impact of the various engineering alternatives. The next chapter will emphasize the environmental compatability of the various engineering alternatives discussed in the siting process.

REFERENCES

1. Calvert, J. and Heilman, W., "A New Approach to PPS," *Proceedings of the ASCE Annual Convention,* 1972.
2. Lees, L., "People, Power, Pollution," California Institute of Technology, September 1, 1971.
3. Porter, A. (Chairman), "Royal Commission on Electric Power Planning," First Report, Royal Commission on Electric Power Planning, Canada, May 1976.
4. Miller, D., "How to Find a Plant Site Without Losing Your Mind," American Electric Power System, 1969.
5. Krishna, P. and Douglas, R., "Nuclear Power Plant Siting Considerations and Experiences," Public Service Electric and Gas Company, 1974.
6. West, L. and Dyar, K., "The Methodology and Criteria Used for a Nuclear Power Plant Siting Study," *Proceedings of the Conference on Nuclear Power Plant Siting,* American Nuclear Society, August 26–28, 1974.
7. Fischer, J. and Ahmed, R., "A Systematic Approach to Evaluate Sites for Nuclear Power Plants," *Proceedings of the Conference on Nuclear Power Plant Siting,* American Nuclear Society, August 26–28, 1974.
8. Geitner, N. and Broome, K., "Systematic Methods for Power Plant Siting Studies," Presented at an American Nuclear Society Power Division Topical Meeting, August 26–28, 1974. Published by Gilbert Commonwealth Associates.
9. Elkins, M. and DiNunno, J., "A Regional Siting for Thermal Power Plants in the State of Ohio," *Proceedings of the Conference on Nuclear Power Plant Siting,* American Nuclear Society, August 26–28, 1974.
10. Coe, R. (Chairman), "Environmental Guidelines," Western Systems Coordinating Council Environmental Committee, December 3, 1971.
11. *Proceedings of the Conference on Nuclear Power Plant Siting,* August 26–28, 1974.
12. *Proceedings of the Seminar on the Engineering Aspects of Siting and Operating Power Plants,* Sponsored by the U.S. Department of Interior and the Atomic Industrial Forum, Inc., 1973. Published by the Atomic Industrial Forum, Inc.
13. Hogerton, J., "Nuclear Reactors," U.S. Nuclear Regulatory Commission, 1970.

14. "Standard Format of Safety Analysis Reports for Nuclear Power Plants," U.S. Nuclear Regulatory Commission, February 1974.

15. Nelson, G., "Water Recycle/Reuse Possibilities: Power Plant Boiler and Cooling Systems," U.S. Environmental Protection Agency, December 1974.

16. Nelson, G., "Technical and Economic Evaluations of Cooling Systems Blowdown Control Techniques," U.S. Environmental Protection Agency, Office of Research and Development, November 1973.

17. "Nuclear Energy Center Site Survey-1975, Part III of V, Technical Considerations," U.S. Nuclear Regulatory Commission, NUREG-0001, January 1976.

18. Fox, C., "Radioactive Wastes," U.S. Nuclear Regulatory Commission, 1969.

19. Dukert, J., "High-level Radioactive Waste: Safe Storage and Ultimate Disposal," U.S. Nuclear Regulatory Commission, 1975.

20. "National Survey on Electric Utilities Security," Burns Security Institute, 1976.

21. Perry, R. and Rutledge, J., "Siting Considerations for Offshore Nuclear Power Plants," Dames and Moore, September 1973.

22. Beck, R., et al, "Underground Nuclear Power Plant Siting," California Institute of Technology, 1972.

23. "Reactions of the Public and Thought Leaders to a Floating Nuclear Electric Generating Plant," Westinghouse Electric Corp., Prepared by Opinion Research Corp., January 1972.

24. "Energy Report," *Spectrum,* Vol. 14, November 7, July 1977.

5

Environmental considerations

In this chapter an attempt is made to present an overview of a few of the numerous impacts associated with the construction and operation of a large power plant. An in-depth presentation of all of the impacts would require several volumes. Therefore, numerous references have been provided for the reader desiring more detailed information.

Chapter 4 presented a summary of the engineering alternatives available in the siting process. Table 5-1 presents a list of some of the environmental factors that can be used in evaluating the various plant configuration alternatives discussed in that chapter. When a particular plant configuration is selected, its environmental impact can be evaluated in more detail. Tables 5-2 and 5-3 present, for the construction phase alone, the type of detail that must be considered in this analysis. The environmental factors that must be evaluated relative

Table 5-1 AEC List of Environmental Factors to be Utilized in Comparing Alternative Power Plant Systems.[1]

Natural Surface Water Body

1. Impingement or entrapment by cooling water intake structure
   a. fish

2. Passage through or retention in cooling systems
   a. phytoplankton and zooplankton
   b. fish

3. Discharge area and thermal plume
   a. water quality, excess heat
   b. water quality, oxygen availability
   c. fish, nonmigratory
   d. fish, migratory
   e. wildlife (including birds, aquatic and amphibious mammals, and reptiles)

4. Chemical effluents
   a. water quality, chemical
   b. fish
   c. wildlife (including birds, aquatic and amphibious mammals, and reptiles)
   d. people

5. Radionuclide discharge to water body
   a. aquatic organisms
   b. people, external
   c. people, ingestion

6. Consumptive use
   a. people
   b. agriculture
   c. industry

7. Plant construction (including site preparation)
   a. water quality, physical
   b. water quality, chemical

8. Other impacts

9. Combined or interactive effects

10. Net effect

Table 5-1 Continued

Ground Water

1. Raising/lowering of ground water levels
   a. people
   b. plants
2. Chemical contamination of ground water (excluding salt)
   a. people
   b. plants
3. Radionuclide contamination of ground water
   a. people
   b. plants and animals
4. Other impacts on ground water

Air

1. Fogging and icing (caused by evaporation and drift)
   a. ground transportation
   b. air transportation
   c. water transportation
   d. plants
2. Chemical discharge to ambient air
   a. air quality, chemical
   b. air quality, odor
3. Radionuclides discharged to ambient air and direct radiation from radioactive materials
   a. people, external
   b. people, ingestion
   c. plants and animals
4. Other impacts on air

Land

1. Site selection
   a. land, amount
2. Construction activities (including site preparation)
   a. people (amenities)
   b. people (accessibility of historical sites)
   c. people (accessibility of archeological sites)
   d. wildlife
   e. land

3. Plant operation
   a. people (amenities)
   b. people (aesthetics)
   c. wildlife
   d. land, flood control
4. Salts discharged from cooling towers
   a. people
   b. plants and animals
   c. property resources
5. Transmission route selection
   a. land, amount
   b. land use and land value
   c. people (aesthetics)
6. Transmission facilities construction
   a. land adjacent to right-of-way
   b. land, erosion
   c. wildlife
   d. flora
7. Transmission line operation
   a. land use
   b. wildlife
8. Other land impacts
9. Combined or interactive effects
10. Net effects

Table 5-2 Potential Environmental Impacts Resulting from Construction Practices.[2]

Construction Phase	Construction Practice	Potential Environmental Impacts
1. Preconstruction	a. Site inventory	Short-term and nominal
	(1) Vehicular traffic	Dust, sediment, and tree injury
	(2) Test pits	Tree root injury, sediment
	b. Environmental monitoring	Negligible if properly done

Table 5-2 Continued

Construction Phase	Construction Practice	Potential Environmental Impacts
	c. Temporary controls	Short-term and nominal
	(1) Stormwater	Vegetation, water quality
	(2) Erosion and sediment	Vegetation, water quality
	(3) Vegetative	Fertilizers in excess
	(4) Dust	Negligible if properly done
2. Site work	a. Clearing and demolition	Short-term
	(1) Clearing	Decrease in the area of protective tree, shrub, and ground covers, stripping of topsoil; increased soil erosion, sedimentation, and stormwater runoff; increased stream water temperatures; modification of stream banks and channels, water quality
	(2) Demolition	Increased dust, noise, solid wastes
	b. Temporary facilities	Long-term
	(1) Shops and storage sheds	Increased surface areas impervious to water infiltration, increased water runoff, petroleum products
	(2) Access roads and parking lots	Increased surface areas impervious to water infiltration, increased water runoff, generation of dust on unpaved areas
	(3) Utility trenches and backfills	Increased visual impacts, soil erosion, and sedimentation for short periods
	(4) Sanitary facilities	Increased visual impacts, solid wastes
	(5) Fences	Barriers to animal migration
	(6) Laydown areas	Visual impacts, increased runoff

Construction Phase	Construction Practice	Potential Environmental Impacts
	(7) Concrete batch plant	Increased visual impacts; disposal of wastewater, increased dust and noise
	(8) Temporary and permanent pest control (termites, weeds, insects)	Nondegradable or slowly degradable pesticides are accumulated by plants and animals, then passed up the food chain to man. Degradable pesticides having short biological half-lives are preferred for use.
	c. Earthwork	Long-term
	(1) Excavation (2) Grading (3) Trenching (4) Soil treatment	Stripping, soil stockpiling, and site grading; increased erosion, sedimentation, and runoff, soil compaction; increase in soil levels of potentially hazardous materials; side effects on living plants and animals; and the incorporation of decomposition products into food chains, water quality
	d. Site drainage	Long-term
	(1) Foundation drainage (2) Dewatering (3) Well points (4) Stream channel relocation	Decrease in the volume of underground water for short and long time periods, increased stream flow volumes and velocities, downstream damages, water quality
	e. Landscaping	
	(1) Temporary seeding (2) Permanent seeding and sodding	Decreased soil erosion and overland flow of stormwater, stabilization of exposed cut and fill slopes, increased water infiltration and underground storage of water, minimize visual impacts

Table 5-2 Continued

Construction Phase	Construction Practice	Potential Environmental Impacts
3. Permanent facilities	a. Transmission lines and heavy traffic areas	Long-term
	(1) Parking lots	Stormwater runoff, petroleum products
	(2) Switchyard	Visual impacts, sediment, runoff
	(3) Railroad spur line	Stormwater runoff
	b. Buildings	Long-term
	(1) Warehouses	Impervious surfaces, stormwater runoff, solid wastes, spillages
	(2) Sanitary waste treatment	Odors, discharges, bacteria, viruses
	(3) Cooling towers	Visual impacts
	c. Related facilities	Long-term
	(1) Reactor intake and discharge channel	Shoreline changes, bottom topography changes, fish migration, benthic fauna changes
	(2) Water supply and treatment	Waste discharges, water quality
	(3) Stormwater drainage	Sediment, water quality
	(4) Wastewater treatment	Sediment, water quality, trace elements
	(5) Dams and impoundments	Dredging, shoreline erosion
	(6) Breakwaters, jetties, etc.	Circulation patterns in the waterway
	(7) Fuel handling equipment	Spillages, fire, and visual impacts
	(8) Oil storage tanks, controls, and piping	Visual impacts
	(9) Conveying systems (cranes, hoists, chutes)	Visual impacts

Construction Phase	Construction Practice	Potential Environmental Impacts
	(10) Waste handling equipment (incinerators, wood chippers, trash compactors)	Noise and visual impacts
	d. Security fencing	Long-term
	(1) Access road	Increased runoff
	(2) Fencing	Barriers to animal movements
4. Project closeout	a. Removal of temporary offices and shops	Short-term
	(1) Demolition	Noise, solid waste, dust
	(2) Relocation	Stormwater, runoff, traffic blockages, soil compaction
	b. Site restoration	Short-term
	(1) Finish grading	Sediment, dust, soil compaction
	(2) Topsoiling	Erosion, sediment
	(3) Fertilizing	Nutrient runoff, water quality
	(4) Sediment controls	Vegetation
	c. Preliminary start-up	Short-term
	(1) Cleaning	Water quality, oils, phosphate
	(2) Flushing	and other nutrients

Table 5-3 Types of Pollutants Resulting from Various Construction Practices.[2]

Construction Phase	Construction Practice	Primary Pollutants
1. Preconstruction	a. Site inventory	Dust, noise, sediments
	b. Environmental monitoring	Visual
	c. Temporary controls	Sediment spoil, nutrients, solid waste
2. Site work	a. Clearing and demolition	Dust, sediment, noise, solid wastes, wood wastes

Table 5-3 Continued

Construction Phase	Construction Practice	Primary Pollutants
	b. Temporary facilities	Gases, odors, fumes, particulates, dust, deicing chemicals, noise, petroleum products, waste water, solid wastes, aerosols, pesticides
	c. Earthwork	Dust, noise, sediment, debris, wood wastes, solid wastes, pesticides, particulates, bituminous products, soil conditioner chemicals
	d. Site drainage	Sediment
	e. Landscaping	Nutrients, pesticides
3. Permanent facilities	a. Transmission lines and traffic areas	Sediment, dust, noise, particulates
	b. Buildings	Solid wastes
	c. Related facilities	Sediment, trace elements, noise, caustic chemical wastes, sediment spoil, flocculents, particulates, fumes, solid wastes
	d. Security fencing and roadway	Sediment, wood wastes
4. Project closeout	a. Removal of temporary offices	Noise, dust, solid wastes
	b. Site restoration	Sediment, dust
	c. Preliminary start-up	Nutrients, petroleum products

to plant operation vary and are a function of the type of plant constructed.

The overview presented here addresses these numerous environmental factors from the perspectives of air, aquatic, terrestrial, social, and special impacts in nuclear plant siting. In discussing the sources and types of these impacts, certain mitigation procedures are also presented. Consequently, the economic trade-offs available for lessening environmental impact through engineering design are an integral part of this discussion.

AIR QUALITY IMPACTS

The numerous considerations involved in investigating air quality impacts may be subdivided into two categories, namely (1) regulatory gaseous emission limits on stack emissions and (2) cooling tower emission effects on the environment.

Stack Emissions

Power plant combustion emissions include sulfur oxides (sulfur dioxide, sulfur trioxide, sulfates, and sulfuric acid mist), nitrogen oxides (nitric oxide and nitrogen dioxide), particulates, hydrocarbons, carbon monoxide, and traces of organic compounds such as aldehydes. Health scientists are most concerned about sulfur dioxide (SO_2), the nitrogen oxides, and particulate emissions. As a result, these emissions have been classified as the primary gaseous pollutants. Table 5-4 presents the National Ambient Air Quality Standards for these pollutants. The detrimental effects of these pollutants can be witnessed in the deterioration of human health, vegetation, structural finishes, and visibility.[1-14]

These pollutants are normally emitted from stacks several hundred feet high. Because stack gases are heated, the exhaust is extremely buoyant. In the presence of atmospheric turbulence the plume content is dispersed and spread over a wide region downwind from the plant. While aloft, interactions can occur between various pollutants, atmospheric oxygen, ozone, dust, and sunlight to intensify the toxicity of the pollutants.[15]

The effects of air pollution may be classified as acute, chronic, and long-term. *Acute effects* are those effects that result from relatively short exposures (less than one month) to high concentrations of pollutants. *Chronic effects* are those effects that occur when organisms are exposed to relatively low-levels of pollutants over extended time periods. *Long-term effects* characterize the abnormal changes that occur in ecosystems over a period of, perhaps, decades.

Table 5-4 National Ambient Air Quality Standards.[3]

Pollutant	Standard		Averaging Time
	Primary	Secondary	
Total suspended particulates	75 μgm/m^3	60 μgm/m^3	Annual geometric mean
	260 μgm/m^3 [a]	150 μgm/m^3	Maximum 24-hr level
Sulfur dioxide	80 μgm/m^3 (.03 ppm)		Annual arithmetic mean
		1300 μgm/m^3 (.5 ppm)	Maximum 3-hr level
	365 μgm/m^3 (.14 ppm)		Maximum 24-hr level
Nitrogen dioxide	100 μgm/m^3 (.05 ppm)	Same	Annual
	250 μgm/m^3 (.13 ppm)	Same	Maximum 24-hr level

[a]Second highest value in any year.

Specific impacts of sulfur dioxide are presented in Table 5-5. Sulfur dioxide is produced when oil or coal is used to fire a power plant. Health effects range from a slight temporary vasoconstriction to a measurable reduction in the elasticity of the lung to thickening of the protective mucus layer over the cilia. Vegetation damage can be characterized by leaf yellowing (chlorosis), transient reductions in live plant biomass, and long-term reductions in crop growth and yield.[17] Finally, if the plume comes in contact with water, an acidic composition can result that will kill aquatic plants, phytoplankton, and/or benthos species due to reduced pH levels.

The majority of the nitrogen oxides that are emitted result from the reaction between two normal constituents of air (oxygen and nitrogen) in the presence of the elevated temperatures emitted from the stacks of fossil-fired plants. Specific impacts are listed in Table 5-6. Nitrogen dioxide (NO_2) is a strong irritant. In chronic exposures evidence of inflammation in humans has been demonstrated. When nitrogen oxides react with water, in much the same manner as sulfur oxides, aquatic organisms are damaged. Another phenomena that produces vegetation damage is nitrogen loading, which causes excessive plant growth.

Table 5-5 Some Effects of Sulfur Dioxide (SO_2) on Human Health and Welfare.[16]

1. HUMAN HEALTH

Concentration (ppm)	Duration or averaging time	Effect	Comments
0.052	24-hr mean	Increased mortality may occur when particulates exceed 6 cohs* soiling index.	Literature from several countries indicates that the main health problems posed by sulfur oxides are related to irritation of the respiratory system. Dose-response curves, based on 1 hr exposures, have been established for some laboratory animals. In laboratory animals and man, sulfuric acid mist and particulate sulfates when present increase the irritant potency of a given SO_2 concentration. The association between long-term exposure to ambient levels of SO_2 and disease incidence rates is conservatively appraised to be intermediate in reliability.
0.25	24-hr mean	Increased death and illness rate may occur when smoke concentration is more than 750 $\mu g/m^3$.	
0.21	24-hr mean	Patients with chronic lung disease may have worsened symptoms if smoke concentrations exceed 300 $\mu g/m^3$.	
0.19	24-hr mean	Increased mortality may occur if high particulate levels are present.	
0.11 -0.19	24-hr mean	Increased hospital admissions for older persons with respiratory diseases.	
0.037 -0.092	Annual mean	When accompanied by smoke concentrations greater than 185 $\mu g/m^3$, increased frequency of lung and respiratory disease may occur.	

*cohs – coefficient of haze: that quantity of particulate material that produces an optical density of 0.01 when measured by light transmission at 400 millimicrons and when compared to the transmission of dust-free filter paper taken as 100%.

Table 5-5 Continued

0.046	Annual mean	When accompanied by smoke concentrations greater than 100 μg/m^3, increased respiratory disease may occur in school children.	

2. DAMAGE TO VEGETATION

Concentration (ppm)	Duration or averaging time	Effect	Comments
0.05 -0.25	4 hr	Moderate to severe injury to sensitive plants when O_3 and NO are also present.	Different species, varieties and even individuals vary in tolerance to SO_2. Over 300 species have been studied. Alfalfa is one of the most sensitive species and is commonly used as a reference.
0.03	Annual mean	Chronic damage and excessive leaf drop may occur.	
0.03	8 hr	Some trees and shrubs show damage.	

3. EFFECT ON MATERIALS

Concentration (ppm)	Duration or averaging time	Effect	Comments
0.12	Ambient mean	Corrosion rates for steel panels may be accelerated by 50%.	Sulfur oxides contribute to damage of many kinds of electrical equipment—including transmission lines. Field observations were substantiated in the laboratory.
0.09	Annual mean	Some dyed fabrics fade.	

4. EFFECT ON VISIBILITY

Concentration (ppm)	Duration or averaging time	Effect	Comments
0.10	Ambient mean	With 50% humidity and the presence of comparable particulate centrations, visibility may be reduced to 5 miles.	Aside from aesthetic considerations, operation of airports can be significantly slowed.

Table 5-6 Some Effects of Nitrogen Dioxide (NO_2) Pollution.[18]

1. HUMAN HEALTH[a]

A. Acute effects due to short-term exposures

Concentration (ppm)	Duration	Dose[b] (ppm sec)	Effect	Comment
0.12	-	-	Olfactory threshold	-
5	10 min.	3,000	Measurable increase in airway resistance.	Transient response.
15 -50	2 hr	108.000 -360,000	Tissue changes in lungs, heart, liver and kidney of monkeys.	Degree of damage proporportional to concentration.
90	30 min.	162,000	Pulmonary edema found 18 hr after exposure —accompanied by a 50% reduction in normal vital capacity.	Accidental occupational exposure.

Table 5-6 Continued

B. Effects due to long-term exposures

Concentration (ppm)	Duration	Dose[b] (ppm sec)	Effect	Comment
0.062–0.109 daily means	Continuous	37,000–66,000 per week	Increased incidence of serious respiratory disease in family groups found during a 6-month study.	Mean level of suspended particulate nitrates during the study was at least 3.8 $\mu g/m^3$.
0.063–0.083	Continuous	38,000–50,000 per week	Frequency of acute bronchitis was increased among infants and school children observed in a 6-month study.	Mean level of suspended particulate nitrates during the study was at least 2.6 $\mu g/m^3$.
5	6–24 hr/day for 3–12 mos.	9,720,000 -158,000,000	Toxicological changes in mice resembling those associated with human emphysema.	Enhanced susceptibility to K. pneumoniae was noticed in mice exposed continuously to this NO_2 level for 3 months.

2. Damage to Vegetation

Concentration (ppm)	Duration	Dose[b] (ppm sec)	Effect	Comment
0.25	Continuous for 8 months	5,200,000	Leaf abscission and decreased yield for naval oranges.	Exposure of beans to NO concentrations of 10 ppm and 4 ppm reduced apparent photosynthesis by 50–70% and 10%, respectively.
0.50	Continuous for 35 days	1,512,000	Leaf abscission and chlorosis on citrus trees.	
1.0	1 day	86,000	Overt leaf injury to NO_2-sensitive plants	

3. OTHER EFFECTS

A. Photochemical oxidant (OX) production

Concentration (ppm)	Duration	Dose[b] (ppm sec)	Effect	Comment
0.04 -0.16	-	-	Maximum daily 1-hr average OX concentration of 0.1 ppm can be associated with these NO_2 levels in the presence of 0.3–1.4 ppm of nonmethane hydrocarbon (6–9 A.M. sunlight).	Pertains to formation of photochemical smog.

B. Stress corrosion

Concentration (ppm)	Duration	Dose[b] (ppm sec)	Effect	Comments
0.066 -0.88 of NO_X	Continuous	-	Nitrogen oxide reaction products have contributed to corrosion and failure of electrical components.	Problem observed in 2 cities during 1965. Average particulate nitrate levels were 3.0–3.4 $\mu g/m^3$.

[a] There is no evidence showing adverse health effects due to nitric oxide (NO) at ambient concentrations.

[b] The concentration-time of exposure product determines nonlethal morbidity effects of NO_2 in toxonomic studies.

Finally, particulates form a significant portion of the combustion product from coal. As noted in Table 5-7, particulates pose a significant threat to human health.

Table 5-7 Some Effects of Airborne Particulates.[19]

1. HUMAN HEALTH

Particulate concentration ($\mu g/m^3$)	Accompanying SO_2	Duration or averaging time	Effect	Comment
750	>715 $\mu g/m^3$	24-hr mean	Excess deaths and increase in illness may occur.	Analyses of numerous epidemiological studies clearly indicate an association between air pollution, as measured by particulate matter accompanied by sulfur dioxide, and health effects of varying severity. This association is most firm for the short-term air pollution episodes.
300	>630 $\mu g/m^3$	24-hr mean	Chronic bronchitis patients likely to have a worsening of symptoms.	To show a small percentage of changes in deaths or increases in hospital admissions associated with coincident higher levels of air pollutants requires extremely large populations. In small cities, these changes are difficult to detect statistically and are most easily demonstrated in major urban areas.
200	>250 $\mu g/m^3$	24-hr mean	Increased absence of industrial workers due to illness.	
100–130	>120 $\mu g/m^3$	Annual mean	Children likely to experience increased incidence of some respiratory diseases.	

Particulate concentration ($\mu g/m^3$)	Accompanying SO_2	Duration or averaging time	Effect	Comment
80–100	Sulfation rate >30 mg/cm^2-mo.	Annual geometric mean	Increased death rates for persons older than 50 yrs is likely	For the large urban communities which are routinely exposed to relatively high levels of pollution, sound statistical analysis can show with confidence the small changes in daily mortality which are associated with fluctuation in pollution concentrations. Such analysis has thus far been attempted only in London and in New York.

2. VISIBILITY

Particulate concentration ($\mu g/m^3$)	Accompanying SO_2	Duration or averaging time	Effect
100–150	-	-	Where large smoke turbidity factors persist, in middle and high latitudes, sunlight may be reduced by 1/3 in summer and 2/3 in winter.

Table 5-7 Continued.

Particulate concentration ($\mu g/m^3$)	Accompanying SO_2	Duration or averaging time	Effect
150	-	-	For particles predominantly in the 0.2μ to 1.0μ size range, and relative humidity less than 70%, visibility can be reduced to as little as 5 miles.

3. MATERIALS

Particulate concentration ($\mu g/m^3$)	Accompanying SO_2	Duration or averaging time	Effect
60–180	Some	Annual geometric mean	In presence of SO_2 and moisture, accelerated corrosion of zinc and steel may occur.

Various emission controls can be used to limit the rate and amount of these pollutants. A scrubber operation can reduce the amount of sulfur expelled. Fluidized bed combustion may be used when coal is used to reduce expelled sulfur. Methods for reducing nitrogen oxides include (1) low-excess air firing (which increases the amount of carbon monoxide expelled due to incomplete combustion), (2) staged combustion, and (3) flue-gas recirculation.[20, 21] Finally, mechanical filters (e.g., cyclone separators or gravity settling chambers), bag-house filters, and electrostatic precipitators can be used to reduce particulate levels. Each control increases the cost of the plant and reduces plant efficiency.

Cooling Tower Impacts

Cooling tower impacts primarily result from the tower plume that can produce drift deposits, fog, and icing. These plumes are often visible. Intensive study of these phenomena have resulted in the development of models that can be used to characterize the size, height, and dispersion of the plume.[22-24]

Drift Deposits. Environmental impacts can take a number of forms. One example is the evaporative cooling tower. If the surrounding air does not have the capacity to absorb excess moisture, some droplets of the circulating water will escape from the tower. These droplets, called drift, carry with them a salt compound of the chemicals present in the makeup water – concentrated several times. In addition, any chemicals added to the makeup water to control biological and chemical reactions in the cooling system will also be present.

The salt is deposited over the area surrounding the tower at a rate dependent upon the tower size, the local meteorological conditions, the amount of drift produced, the drift droplet size distribution, and the salinity of the water. The amount of drift emitted from a modern cooling tower can be on the order of 0.0002% of the circulating water. Upon being deposited, the salt can damage vegetation through direct contact and indirectly change soil properties. Weathering and corrosion of metals may also occur. The chlorine content of ground water may also be increased. For these reasons, predictions of drift and salt impacts are extremely important in power plant design and siting.

Fog. Fog is essentially a cloud that has formed at ground level. The conditions that promote water vapor condensation in ground-level air can lead to fog conditions with the proper ambient relative humidity when the cooling tower water is added to the surrounding ambient air. Depending upon relative atmospheric stability and wind velocity, this fog can concentrate in a dense pattern over a small area or be dispersed to form a sparse pattern over a wider region.

Visible Plumes. In addition to causing ground-level fog, cooling towers can produce visible plumes. When the entrained ambient air produces a mixture above the saturation level of the air, visible plumes normally result. Cold weather and high humidity are the atmospheric conditions most conducive to long visible plumes. Under normal circumstances, a plume usually presents only an aesthetic problem. However, in the presence of high humidity, cold weather, and high wind speed, the plume may be brought down to ground relatively close to the source.

Icing. Under high humidity, high wind speed, and freezing temperatures, icing of roads and structures can result. Even when not visible, contact of a moist plume with a colder surface can result in heat transfer to the surface and condensation of the moisture. Under freezing conditions ice will be formed. As the plume moves further away from the cooling tower, however, the possibility of icing occurring decreases.

AQUATIC IMPACTS

Utilization of open cooling systems involves the withdrawal, treatment, and discharge of water. When a river, estuary, or lake is used as the water source, these functions can affect the welfare of organisms and can cause biological stress. A few of the primary impacts upon water quality from this process are thermal pollution, chemical toxicity, entrainment, entrapment, and dissolved oxygen sag. Each of these impacts are treated briefly in the following sections.

A total aquatic community, as indicated in Figure 5-1, can be divided into three major elements. The microscopic organisms (bacteria, algae, and plankton), represent the base of the community. The macroinvertebrates, or so-called "bottom organisms," (worms, snails, insects, and their larvae), form the middle group. Finally, fish form the top group in the community. A knowledge of specific projected plant parameters and the biological characteristics of the various sectors of the aquatic community is required before an accurate assessment of biological impacts can be accomplished.

Thermal Effects

The dispersion of thermal discharges has been studied for many years. The consensus of most of these studies is that the benthos,

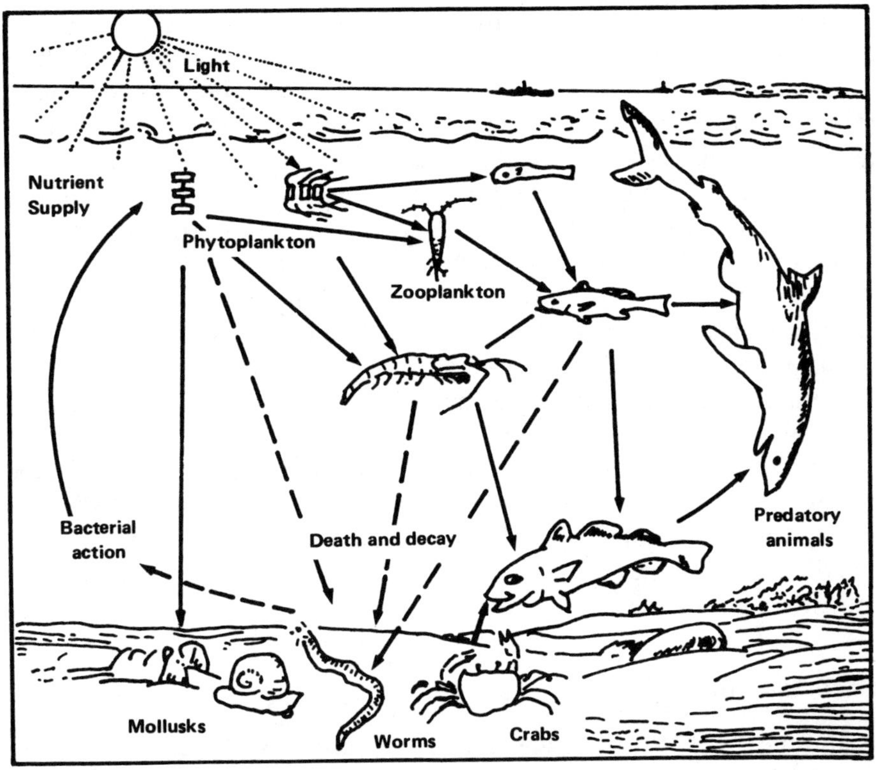

Figure 5-1 The aquatic environment.[25]

bottom dwellers, are the most prone to heat-plume impact because of their immobility and constant exposure. While those organisms directly in the discharge and in the zone of discharge mixing may be subjected to drastic and rapid thermal consequences, those organisms beyond the mixing zone are seldom adversely affected. Another consequence is that migratory species usually do not return.

The water temperature in the region of discharge can produce lasting effects on aquatic life. Fish, for example, become sluggish at higher temperatures and have difficulty in obtaining food.[26] Given time, most organisms will adapt to temperatures slightly different from those to which they are accustomed. Higher temperatures can usually be tolerated easier than lower temperatures. If, however, temperatures reach an uncomfortably high stage, fish will often migrate to cooler waters. It is important to note that large fish kills usually occur in the winter when a forced outage closes down a generating plant thereby stopping the heated effluent. Such sudden temperature drops are known as "thermal shock."[6]

The effects of elevated temperature include (1) shifts in the ecosystem population structure, (2) a reduction in the number of species, and (3) an increase in the toxic action of pollutants (by increasing the metabolic rate of the organism and by speeding the chemical reaction of pollutants). In addition, these temperatures can act as depressants and restrictors.

To minimize these adverse effects, discharge designs should include optimal mixing, minimization of the rate of temperature changes, and heat diffusers.[27] While conventional shoreline surface discharge systems are least expensive, they do not provide protection to the biologically productive shoreline area from the high temperatures in the discharge plume. An alternative is to employ offshore, submerged, multiport diffusers.

Chemical Toxicity

A number of substances are normally present in the circulating water of a cooling system. Chemical treatment is needed to counter adverse effects of these substances. The system, then, contains the intake water, various adverse substances, and chemicals. The major problems that result are scale formation, fouling, microbiological growth, and corrosion.[25]

In most areas of the country, cooling system makeup water is taken from wells, rivers, or lakes. When this water is alkaline, the alkalinity must be removed. With acidic makeup water, lime is used to neutralize the pH. A large number of chemicals are available for water treatment (see Table 5-8). Care must be taken, however, to select chemicals that do not interact in adverse manners.[28]

In power plant cooling systems, the condenser is the most vulnerable piece of equipment because high tube wall temperatures are present and because these tube walls must be protected to avoid loss of heat transfer capability. *Corrosion,* the electrochemical deterioration of metal, is one action that must be avoided. *Scaling,* the depositing

Table 5-8 Some Chemicals Associated with Power Plants.[25]

1. Corrosion and Scale Inhibitors

Chromates

Sodium chromate
Na_2CrO_4

Sodium dichromate
$Na_2Cr_2O_7 \cdot 2H_2O$

Zinc chromate
$ZnCrO_4$ or $ZnCr_2O_4$

Zinc dichromate
$ZnCr_2O_7 \cdot 3H_2O$

Potassium chromate
K_2CrO_4

Potassium dichromate
$K_2Cr_2O_7$

Phosphates and Polyphosphates

Calcium metaphosphate
$Ca(PO_3)_2$

Sodium phosphate
Na_3PO_4

Sodium metaphosphate
$(NaPO_3)_3 \cdot 6H_2O$

Sodium hexametaphosphate
$(NaPO_3)_6$

Sodium tripolyphosphate
$Na_5P_3O_{10}$

Sodium pyrophosphate
$Na_4P_2O_7$ or $Na_4P_2O_7 \cdot 10H_2O$

Zinc phosphate
$Zn_3(PO_4)_2$

- Sodium orthophosphate
  $Na_3PO_4 \cdot 10H_2O$
- Calcium phosphate
  $CaPO_4$

Organic polyphosphates (proprietary, organic-based compounds)

Glassy Silicates

Sodium silicate
$Na_2O \cdot XSiO_2$ (where x = 3–5)

Nitrites and Nitrates

Sodium nitrite
$NaNO_2$ (rare use)

Sodium nitrate
$NaNO_3$

Potassium nitrate
KNO_3

Cyanates

Sodium ferrocyanate
$Na_4Fe(CN)_6 \cdot 10H_2O$ (rare use)

Fluorides

Sodium fluoride
NaF (rare use)

Amines (also used as biocides)

Octadecylamine

Ethylenediamine

Cyclohexylamine

Benzylamine

Morpholine

Chelating Agents

Ethylenediamine tetraacetic acid (EDTA)

Table 5-8 Continued.

Nitrilotriacetic acid (NTA)

LTSR – "low temperature scale remover" (a proprietary compound produced by Dow Chemical)

2. Corrosion Products

Copper ions
Cu^{2+}

Zinc ions
Zn^{2+}

Inorganic Scale and Precipitates

(generally insoluble in water)

Calcium carbonate
$CaCO_3$

Calcium phosphate
$CaPO_4$

Calcium sulfate
$CaSO_4$

Calcium hydroxide
$Ca(OH)_2$

Magnesium carbonate
$MgCO_3$

Magnesium hydroxide
$Mg(OH)_2$

Magnesium phosphate
$Mg_3(PO_4)_2$

Iron oxides
Fe_2O_3 and Fe_3O_4

3. Cleaning and Neutralizing Compounds

Alkaline Cleaning Stage

Sodium hydroxide
$NaOH$

- Sodium sulfate
  Na_2SO_4

Calcium hydroxide
$Ca(OH)_2$

Sodium phosphate
Na_3PO_4

Alkaline Cleaning Stage

Sodium triphosphate
$Na_5P_3O_{10}$

Ammonium hydroxide
NH_4OH (rare use)

Acid Cleaning Stage

Citric acid
$HOC(CH_2CO_2H)_2CO_2H$

Sulfuric acid
H_2SO_4

Neutralizing (Passivating) Stage

Sodium carbonate
Na_2CO_3

Sodium sulfate
Na_2SO_4

Sodium phosphate
Na_3PO_4

Sodium diphosphate

Sulfuric acid
H_2SO_4

Lithium hydroxide
$LiOH$

- Lithium
  Li

Morpholine
C_4H_9NO

Sodium lignosulfonate

Cyclohexylamine
$C_6H_{11}NH_4$

Ammonium sulfate
$(NH_4)_2SO_4$

Ammonium hydroxide
NH_4OH

Ammonia
NH_3

Oxygen Reducers

Hydrazine
N_2H_4 or NH_2NH_2

Morpholine
C_4H_9NO (rare use for this purpose)

Sodium sulfite
Na_2SO_3

Cobalt sulfate
$CoSO_4$ (catalyst for Na_2SO_3)

Cobalt
Co

Reactivity Control

Boric acid
HBO_2 (meta) and
H_3BO_3 (ortho)

Boron
B

4. Biocides (particularly cooling tower use)

Oxidizing Biocides

Chlorine
Cl_2

Bromine
Br_2
(rare use)

Sodium hypochlorite
NaOCL

Calcium hypochlorite
$Ca(OCl)_2$

Potassium permanganate
$KMnO_4$ (rare use)

Chlorinated cyanurates and inocyanurates

Persulfate Compounds

Potassium hydrogen persulfate

Nonoxidizing Biocides

1. Chlorinated and/or phenylated phenols:
   a. Chloro-o-phenylphenol
   b. 2-tert-butyl-4-chloro-5-methylphenol
   c. o-benzyl-p-chlorophenol
   d. 4, 6-dichlorophenol
   e. 2, 4-dinitrochlorobenzene
   f. 1, 6-dinitrochlorobenzene
   g. 2, 4, 5-trichlorophenol
   h. 1, 3-dichloro-5, 5-dimethylhydranotin
   i. Trichloromethyl sulfone (Bis)

   Sodium salts (ates) of:

   o-phenylphenol

   2, 4, 5-trichlorophenol
   (sodium 2, 4, 5-trichlorophenate)
   (widely used)

   Chloro-2, phenylphenol

   2-chloro-4-phenylphenol

   2-bromo-4-phenylphenol

   2, 3, 4, 6-tetrachlorophenol

   Pentachlorophenol
   (Sodium pentachlorophenate)

Table 5-8 Continued

Potassium salts (ates) of:

2, 4, 5-trichlorophenol

2. Quaternary amines (quaternary ammonium compounds)

Dilauryl dimethyl ammonium chloride

Dilauryl dimethyl ammonium oleate

Dodecyl trimethyl ammonium chloride

Trimethyl ammonium chloride

Octadecyl trimethyl ammonium chloride

N-Alkyl benzyl-N,N,N-trimethyl ammonium chloride

Alkyl-9-methyl benzyl ammonium chloride

Lactoxy mercuriphenyl ammonium lactate

Alkyl dimethyl benzyl ammonium chloride

3, 4-dichloro benzyl ammonium chloride

Phenylmercuric trihydroxy-ethyl ammonium lactate

Phenylmercuric triethanol ammonium lactate

Alkyl (C_{12} to C_{18}) dimethyl benzyl-ammonium chlorides

1-alkyl (C_6 to C_{18}) amino-3-aminopropane monoacetate

3. Organo-metallic Compounds

Organotins

Bis (tributyl tin) oxide

Organosulfurs

Disulfides

Organothiocyanates

Methylene bisthiocyanate

4. Cationic Surface Active Agents

Sulfonium

Phosphonium

Arsonium

Iodonium

5. Dithiocarbamic Acid Salts

Sodium dimethyl diethyl dithio-carbamate

Disodium ethylene bisdithiocarbamate

6. Organic Amines (often used with pentachlorophenol)

Primary rosin amines

Sodium carboxyethyl rosin amine

Rosin amine acetate

Other amines (primary beta-amines and beta-diamines)

Chloroamine

Benzylamine

Cyclohexylamine

Ethylenediamine

Polyethyleneamine

7. Zinc and Copper Salts	8. Acrolein
Zinc sulfate	CH_2:CHCHO
• Zn^{2+} ion	9. Arsenates
Copper sulfate	Arsenic acid
• Cu^{2+} ion	Sodium arsenite
copper citrate	$NaAsO_2$

of materials on surfaces, is another action which must be avoided. In addition to introducing anticorrosion chemicals and antifoulants, chlorine is also added to control the growth of algae, fungi, and bacteria.

While it is well known that chlorine residuals are harmful to aquatic life, there are many other chemicals utilized in the treatment of cooling water that may be considered to be toxic. Consequently, a well-designed power plant will include a well-instrumented chemical control system on its cooling system. Such a design, however, will have an impact on the cost of the plant.

Entrapment

Entrapment, often called impingement, is caused by hydraulic forces in the stream of a power plant prior to the flow of the stream through the screens.[15] These forces gather aquatic life and restrict movement. For most species this action is lethal. Death can occur due to starvation, exhaustion, or asphyxiation. (Asphyxiation can occur when the aquatic life is forced against the screen by velocity forces that prevent proper gill movement or is sprayed by the descaling screen wash or is removed from water for a prolonged period.)[29]

The impingement effect can be significantly influenced by the location, design, and capacity of the intake structures. The various components of the intake structure include screening systems, trash racks, pumps, and fishhandling and bypass equipment. Selection of (1) an appropriate screen mesh size, (2) air bubble screens to keep fish away from intake structures, (3) a fish bypass area or return channel, so captured fish are returned downstream, (4) light barriers, (5) sound barriers, (6) inclined screens, or (7) a combination of these approaches can reduce the impingement effect.

Entrainment

Eggs and newly hatched fish larvae are planktonic – they drift with the current. In the presence of power plant water intake currents, eggs and larvae are entrained into the plant with the intake water. During their passage through the cooling system, entrained organisms are subjected to thermal stresses produced by temperature changes, mechanical stresses produced by the shear and pressure forces in the cooling system, and chemical stresses produced by the chemicals in the makeup water.

Organisms are subjected to temperature variations ranging from rises in the condenser to drops in the cooling system. This fluctuation can be lethal. A common term used to express this thermal stress is the "dose" – the product of temperature rise and duration of exposure. High velocity discharges, such as once-through cooling systems, rapidly dilute the effluent and reduce the dose. Cooling towers and plants, having long effluent canals, retain the heated water longer and thus increase the dose.

Other Impacts

The dredging and filling of intake structure foundations and barge canals can affect bottom dwellers and can temporarily increase water turbidity. Construction site runoff can also increase water turbidity and create erosion problems.

One impact of note results from thermal changes. These changes can produce a sag in the dissolved oxygen, thereby permitting excessive biota growth and increased metabolic rates. Another concern relative to dissolved oxygen is the flow rate of water near dams. Constant regulation of flow is needed to prevent excessive fish kills.

While there are other impacts associated with water quality, those discussed above are the most important in power plant siting. Attention will now be turned to terrestrial and social impacts.

TERRESTRIAL IMPACTS

Most of the impacts that occur to the terrestrial ecology are incurred during construction. The impacts during operation, such as airborne effluents and chemical toxicity, have already been addressed.

Construction impacts can include damage to vegetation, decline of wildlife, reduction of agricultural resources, and impairment to conservation and natural areas.

An accurate assessment of terrestrial impacts will, by nature, include consideration of (1) effects on flora and fauna populations and interactions, (2) rare and endangered species and habitats, (3) hunting and agricultural resources, and (4) sensitive conservation and natural areas. Disturbance due to construction will range from complete destruction of all plant and some animal life at the actual site of construction to relatively mild disturbances of the biota and physical features of the surrounding area. These changes in physical features will result in an altered ecological community structure. Changes in the community structure can result in changes in productivity patterns. Normally these changes take the form of reductions in species populations and reproduction rates.[8]

While species that are prevalent throughout the year will be affected the most, migratory species should not be overlooked. Sensitive habitat components that must be analyzed are (1) breeding areas, (2) unique vegetation required for animal health, (3) unique effects of plants or animals on the life cycle of other plants or animals, and (4) unique features of the area that attract animals.

Since terrestrial impacts will be fairly localized, this impact analysis area should be required for each site. To ensure that rare or endangered species are not affected, collected data should include the following information:

1. acres of woodland that would be lost or damaged – by type and quantity,
2. acres of wetlands that would be lost or damaged – by type and quantity,
3. acres of wildlife habitat that would be lost or damaged – by type and quantity, and
4. acres of agricultural land that would be lost or damaged – by type and quantity.

The impact analysis should address its attention to

1. the uniqueness of the habitat,
2. the degree to which rare or endangered species or habitats would be affected,

3. the degree to which biotic communities would be damaged by air pollutants from the power plant,
4. the degree to which biotic communities would be damaged by gaseous radionuclide emissions from the power plant,
5. the degree to which migration routes of water fowl and game animals would be affected, and
6. the degree to which human resources would be affected.

SOCIAL IMPACTS

This heading includes a wide variety.[30-36] Basically, the impacts revolve around (1) competition for land and water, (2) availability of community services, and (3) cultural threats.[37] During the five to seven year construction period presently required for modern power plants, most of the major features of a characteristically rural population will be affected.[38] Recent studies have attempted to clarify the distinct social issues involved in power plant siting.[39-41]

Table 5-9 depicts the major social impacts associated with power plant construction as evidenced by a recent study. The sections that follow address the more important categories.

Public Service[30, 37]

The rapid influx of a large construction force into an area produces a significant impact on both public services (i.e., education, fire protection, and police protection) and public facilities (i.e., telephone service, water service, sewage treatment, and roads). More often than not, a community will be unprepared for rapid growth, and services suffer. Even worse, the rapid population decline that follows plant construction often results in an overabundance of available services with an inadequate revenue base to support them.

The ability of a community to expand its public services involves the size of the community, the extent to which present services are being utilized, and the ability to finance new services. The financing problem is usually solved by tax payments from the utility. If the increased tax base does not provide sufficient revenue, the tax rate structure can be altered to accommodate increased taxes. In addition, for communities having a planned pattern of growth, application

Table 5-9 Major Social and Economic Impacts Associated with Power Plants.[37]

Note: The main headings describe the major sectors of social and community life which could potentially be impacted; the subheadings describe the types of changes which might take place in each sector.

1. Relocation Impacts
   A. Displacement of residents from the site area (These people could be confronted with a new housing market of escalating prices. Prices may be higher than the compensation they received for their property.)
2. Demographic Impacts
   A. Rapid population growth caused by in-migration of the construction force workers employed in service or support jobs and those seeking employment who may remain unemployed
   B. Change in the ethnic composition of the population
   C. Change in the sex ratio of the population
   D. Change in the age structure of the population
   E. Change in the educational status of the population
   F. Change in the population density of the area
   G. Change in the urbanization rate of the area
3. Local Economic Impacts and Taxes
   A. New job opportunities in the community
   B. Change in the total median family income in the area
   C. Change in the employment rate in the area
   D. Change in the retail prices in the area
   E. Demand increases for retail services
   F. Change in the assessed value of property in the area
   G. Change in prevailing real estate prices
   H. Change in the cost of living in the area
   I. Change in the budget and property tax levy review
   J. Increased revenues from the sales tax
   K. Change in the community's gross product
   L. Change in the business and industrial composition of the community
   M. Change in rates of business starts and failures
   N. Change in economic base
   O. Increased economic planning required
4. Public Works and Transportation Impacts
   A. Water
      1. Demand increases for major transmission and distribution system for water supply
      2. Need increases to develop new drinking water sources

Table 5-9 Continued.

- 3. Need to provide more technology to protect water quality
- 4. Need to provide storm runoff systems as urbanization increases
- 5. Change in the demand for water

B. Sewage Impacts
- 1. Increased demand for sewer collection treatment facilities
- 2. Increased need for facilities for sludge disposal and recycling
- 3. Failure of subsurface sewage system

C. Solid Waste Impacts
- 1. Increased demand on collection and storage of solid waste
- 2. Increased demand on disposal and recycling facilities
- 3. Increased inspection of handling and disposal facilities
- 4. If land disposal, is there adequate land and suitable conditions?

D. Air Pollution Impacts
- 1. Increased need for air pollution control policies
- 2. Increased need to enforce existing policies
- 3. Increase in dust and air matter from trucks going to the site

E. Noise Pollution
- 1. Increased need for policies to control disruptive construction and traffic noise
- 2. Increased need to enforce existing noise ordinances

F. Transportation Impacts
- 1. Increased demand for a variety of services
  - a. highways and expressways
  - b. principal, minor arterials and collectors
  - c. public parking facilities
  - d. traffic control and safety
  - e. public transit
  - f. airport, land and harbor facilities
  - g. bikeways
  - h. rail transport
- 2. Increased need to maintain highways, roads, and major arterials
- 3. Increased demand for special transportation services, e.g., construction workers
- 4. Increased need for transportation planning

5. Land Use Impacts – Agricultural/Residential/Commercial/Industrial/Forest /Recreational

A. Change in existing land use patterns (from agricultural to more urbanized uses)

B. Increased demand on services provided by planning and engineering departments: zoning, subdivision regulations, comprehensive planning, building permits and inspections, and code enforcement
C. Interference with present land use activity nodes
D. More sophisticated management tools required

6. Housing Impacts
   A. Changes in residency duration and mobility
   B. Increased demand for housing of various types: rental units, mobile homes, rooms in houses, single family dwellings
   C. Increase in housing values and rental costs
   D. Changes in ethnic and social class segregation in housing
   E. Increased alterations from existing housing market structure – new units, size, price range
   F. Increased possibility of out-migration of existing residents
   G. Increased demand for low income housing

7. Recreation/Culture Impacts
   A. Increased demand for a variety of recreational facilities and services, public/private, indoor/outdoor
      1. Ballparks and playgrounds
      2. Tennis courts
      3. Swimming pools
      4. Parks
      5. Taverns
      6. Theaters
   B. Increased demand for the community to plan, acquire or construct, maintain, or operate recreational facilities
   C. Increased need to organize and direct recreational programs
   D. Increased demand on cultural facilities
      1. Museums
      2. Libraries
      3. Cultural centers (coliseums, auditoriums, historical sites, fairgrounds)
   E. Increased demand for neighborhood centers and programs

8. Public Safety Impacts
   A. Increased demand/need for basic police and fire services – apprehension and detection, crime prevention, fire protection
   B. Increased demand on police enforcement and fire protection support systems – communications, equipment maintenance training
   C. Increased demand/need for legal and judicial services

Table 5-9 Continued

- D. Extent and quality of corrections facilities and services (drug, alcohol, delinquency, criminal)
- E. Inadequate jail facilities

9. Educational Impacts
   - A. Increased demand and need for facilities and personnel in primary and secondary education
   - B. Increased strains on the budget and quality of the educational system

10. Physical and Mental Health Service Impacts
    - A. Increased demand for professional services and facilities
    - B. Increased strains on the budgets and quality of physical and mental health care services
    - C. Increased demands on laboratory services
    - D. Increased demands on a variety of services
      1. Family planning
      2. Prenatal care
      3. Chronic diseases
      4. Home health care
      5. Emergency services, e.g., ambulances

11. Public Health Impacts
    - A. Increased demands on sanitation inspections – restaurants, school buildings, water sampling, day care, foster homes, etc.
    - B. Increased demand for public health programs

12. Social Services Impacts
    - A. Increased demand on government agencies for various types of social service payments: welfare, unemployment, food stamps
    - B. Increased demand/need for various services:
      1. Alcohol abuse programs
      2. Interpersonal counseling
      3. Drug abuse programs
      4. Child welfare services
      5. Manpower employment-job training, work experience programs

13. Social and Psychological Well-being Impacts
    - A. Change in the type and number of crimes in the community
    - B. Change in types and numbers of social problems – divorce, alcoholism, suicide
    - C. Change in perceived quality of life
    - D. Change in life-styles as a result of urbanization – accelerated way of life
    - E. Possible conflicts between newcomers and oldtimers

14. Community Structure Impacts
    A. Changes in the number, types of community organizations
    B. Changes in the participation rates in community organizations
    C. Rise in ethnic organizations and activities
    D. Increased demand on local mass media – newspapers, radio, television stations
    E. Replacement of traditional rural community values and preferences by urban values and preferences, e.g., dramatic increase in fast food and apartment-oriented services
15. Organizational
    A. Changes in family living patterns (single, couples, families, etc.)
    B. Changes in informal social interaction
    C. Changes in the number and types of churches
    D. Changes in the socio-economic class structure
16. Political Impacts
    A. Changes in the strengths of local political parties
    B. Changes in citizen participation in politics – voting, communicating, acting
    C. Changes in the strength and structure of informal community power systems
    D. Changes in relationships between local, state, and federal governments
    E. Increased demands on local government budget

for financial support can be made to several Department of Housing and Urban Development programs.

Population Impacts

Employment of local labor can mitigate the impact of a large construction force. In most cases, however, more than half of the labor force is composed of critical skills that must be imported from other communities. In these instances, the natural tendency is for the personnel to relocate their families. Changes in population mix in the critical areas of sex, age, race, and educational background may cause local resentment. This problem area has caused recurring intervenor activity in some energy projects in the West.

In addition to the competition for jobs, disparities in income, and new psychological stress on residents, the increase in population often causes housing shortages. Low cost mobile home parks accrue. Rental property increases as a function of supply and demand. Rapid construction of low quality dwellings results. Each of these housing categories can serve as a new revenue source for the tax base. The major thrust for local government legislative activity, however, usually revolves about new housing and construction controls.

Land Use Impacts[42]

Both the physical construction of the plant and the changing social structure affect land use patterns. Agricultural land is converted to residential property. Urban oriented community development projects are initiated in rural areas. Zoning structures change. Building permit and code enforcements deviate under pressures for new construction.

Land use conflicts are especially acute in coastal zone areas and along rivers as power plants compete with other industrial developments. The competition for water rights is even more critical, especially in the West. In this region of the country it is not unusual for a utility to purchase large tracts of agricultural land in order that the water rights associated with the land might be gained. The water, once used for irrigation, is used for cooling. The land, once tilled, is left fallow.

Health Impacts

New power plants indirectly create an increased demand for many health services. In addition to requirements for increased sanitation services, new medical services (i.e., hospital, emergency room, prenatal care, and pharmaceutical) are required. Public health programs, including drug abuse, alcohol abuse, child welfare, and psychological counseling programs, have to be developed. Human resource services have to be initiated to address problems of divorce, unwed mothers, alcoholism, and suicide.

A recent study has set forth several recommendations to improve investigation and mitigation of impacts in the sociological area.[37] First, the community's ability to manage expected impacts should be established. Second, the community should develop a plan for meeting the impacts of the new power plant. Third, the state should interact with the community by providing resource support where needed. Finally, joint state/local planning should be initiated to facilitate a response to new unexpected impacts unique to the community. With such an approach social impacts associated with power plant siting can be mitigated.

NUCLEAR PLANT IMPACTS

The most important concern in siting nuclear power plants is the potential for radiological hazard to the surrounding population. Aside from the normal siting impacts already mentioned, this aspect of radiological exposure adds another dimension to the siting process.[43–48] Risk to the public can be minimized by intense safety design and by redundant systems that control radioactive release from either normal operation or accidents. Figures 5-2 and 5-3 illustrate the radioactive pathway to humans and other organisms in the event of release.

Constraints

By federal law, an exclusion zone must be maintained around a plant site in order to ensure the safety of the public under normal operation. Systems must be present to collect and dispose of the various solid, liquid, and gaseous radioactive wastes developed during system operation.[50–52] Also, normal operations must be monitored and meet detailed technical specifications so that radiation releases are held to the lowest practical level.

Another constraint placed on nuclear facilities relates to population densities. Specifically, the proximity of a nuclear plant to a major metropolitan area is limited. Again, this constraint is intended to minimize the effects of normal operation or any abnormal occurence.

Radioactive Wastes

There are three actions that result in radioactive wastes. First, fission products can be released into the primary coolant because of a defective fuel element. Second, fission products can originate in the primary coolant as a result of the fissioning of "tramp" uranium on fuel

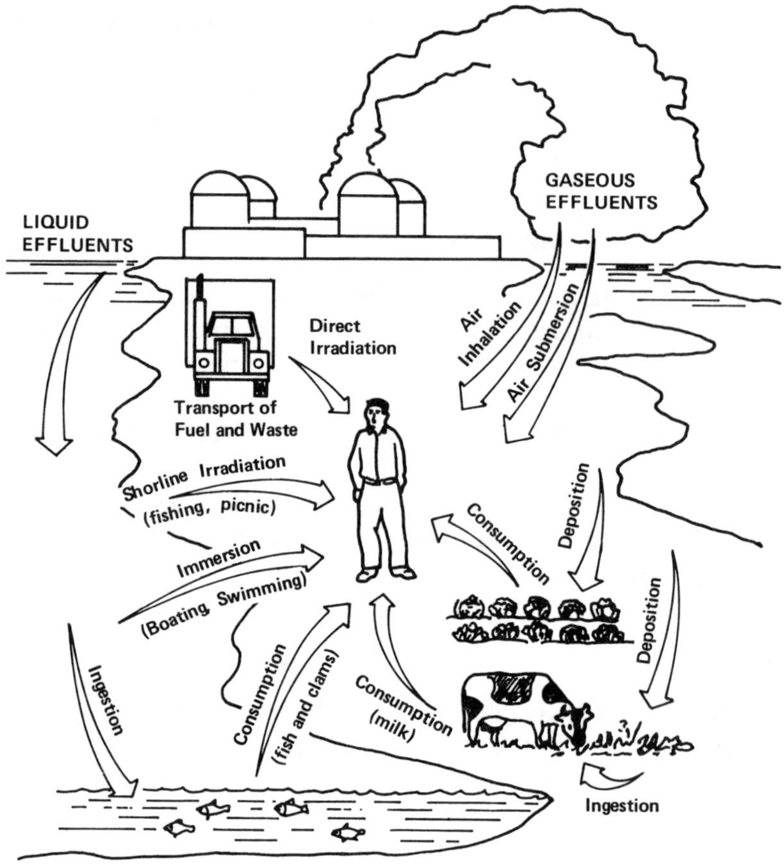

Figure 5-2 Man's radiation exposure pathways.[49]

surfaces. Finally, activation of corrosion products, chemical additives, and the coolant itself can occur.

The gaseous effluents consist of the fission product noble gases (krypton and xenon) and halogens (primarily iodine), tritium (in water vapor form), and particulate matter (fission products, their daughters, and activated corrosion products). Approximately fifty fusion and activation products need to be considered in the liquid effluent. Solid wastes can result from processing liquid streams in the form of

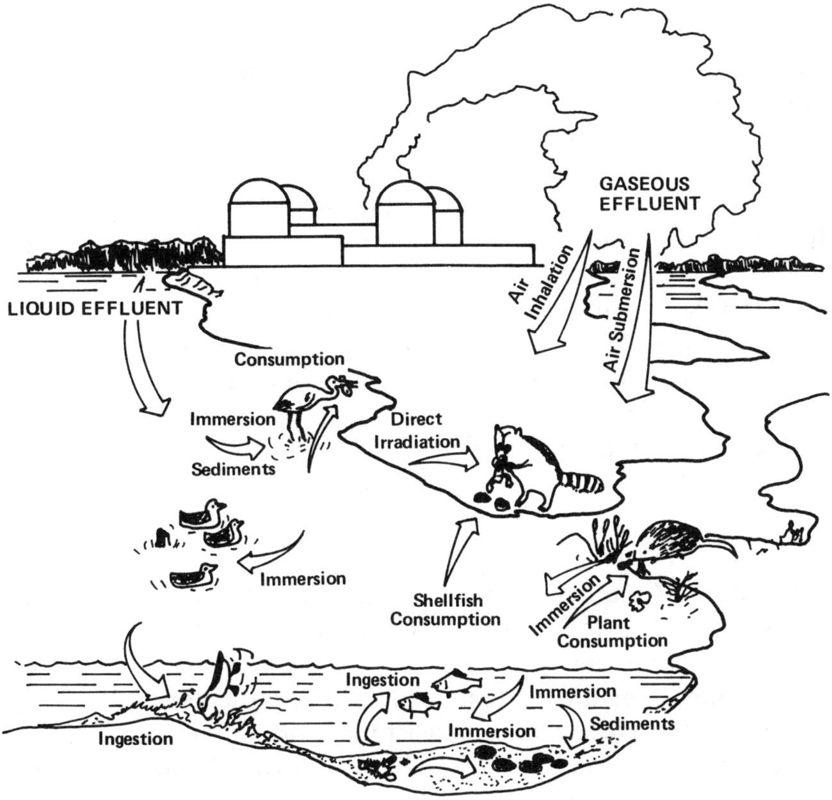

Figure 5-3 Radiation exposure pathways for organisms other than man.[49]

1. spent demineralizer resins,
2. evaporation bottoms,
3. sludge tank residues, or
4. filter trappings

and contaminated objects such as

1. rags,
2. clothing,
3. paper,
4. laboratory materials, or
5. tools.[53, 54]

Leaks

Leaks in the gaseous system can originate in the reactor building, the radwaste building, and the turbine building. Leaks in the liquid system can occur from the reactor coolant, the shim bleed, and several auxiliary systems.

With respect to leaks, three points must be brought into focus. First, the NRC requires that all radioactive effluents be monitored. Second, the NRC licensing provisions cover design production and release levels of radwaste as well as radiation dose levels. Third, the average annual dose from normal operation of nuclear plants has been very small. (Data on annual whole-body radiation doses are presented in Figure 5-4.) Because of the stringent NRC requirements, it is apparent that appreciable leaks have been almost nonexistent.

Radiation Doses

Under the NRC guidelines for limiting radiation doses to the public, levels must be "as low as practical" from nuclear power plant operation. The maximum doses that may be received by the most exposed members of the general public are specified and are presented in Table 5-10. These values constitute approximately a 5% increase in the total radiation exposure that an individual can receive exclusive of medical exposures. The risk of cancer to any individual actually receiving these doses is given in Table 5-11. The risk of generic damage to the exposed individual is given in Table 5-12.

Irradiation and Risk Factors

Beside the possible impacts listed above from normal operations, there is the additional possibility of risk by inadvertent irradiation due to a nuclear accident. Since there is no historical evidence available upon which to postulate this potential, the number of accidents, or related deaths, is calculated using the much publicized Rasmussen Report.[7]

This NRC sponsored reactor safety study extended "fault-tree" analysis techniques to make actual risk estimates for nuclear reactor accidents. Fault-tree analysis uses failure rates of system components,

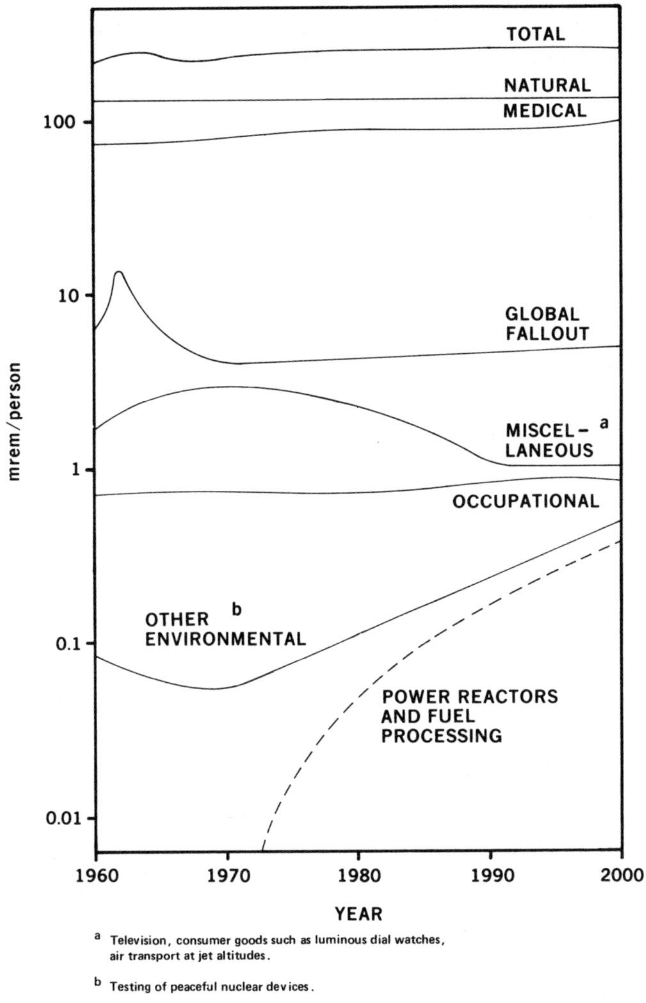

Figure 5-4 Estimated average whole-body radiation doses in the United States.[15]

taken from power plant operation data, to compute the probability of a failure for a system built with those components.

The study verified that the major contribution to overall risk from nuclear power plants is an accident where the reactor core melts due to the failure of all cooling systems. The risk of death from such an accident was estimated to be extremely small – smaller than many risk values which historically have been neglected as inconsequential.

Table 5-10 NRC Guidelines for Limiting Radiation Doses to the Public.[15]

Method of Exposure	Organ Exposed	Design Basis Maximum Dose
Immersion in and ingestion of aqueous discharges	Whole body[a]	5 mrem
Immersion in and ingestion of gaseous discharges	Whole body[a]	5 mrem
All iodine releases	Thyroid gland	15 mrem

[a]A single individual is not expected to receive the maximum dose from both aqueous and gaseous plant discharges.

Table 5-11 Risks (Accumulated Annually) to Individuals (and Descendants of Individuals) Exposed to the NRC Radiation Dose Limits of Table 5-10.[15]

	Additional Risk Dut to Radiation	Normal Risk Level
Risk of cancer (assuming 10 mrem)[a]	$\frac{2}{1,000,000}$	$\frac{1,520}{1,000,000}$
Risk of thyroid nodule (assuming 15 mrem)	$\frac{3}{1,000,000}$	$\frac{\sim 10,000}{1,000,000}$
Risk of genetic defect in children (assuming 10 mrem)[a]	$\frac{0.07 \text{ to } 1.4}{1,000,000}$	$\frac{60,000}{1,000,000}$
Risk of genetic defect in any descendants (assuming 10 mrem)[a]	$\frac{0.4 \text{ to } 10}{1,000,000}$	$\frac{60,000}{1,000,000}$

[a]Assuming the individual receives maximum dose from both the gaseous and aqueous releases (contrary to NRC assumptions).

Table 5-12 Risks of Genetic Defect Induction from Radiation.[12]

A. Specific Genetic Damage: Effects per million live-born descendants due to 1-rem radiation dose to all individuals in one generation

Effect	Cases in First Generation	Total Cases Over all Future Generations[a]	Current Incidence per Million Live Births
Autosomal dominant traits	10-100	50-500	10,000
X-chromosome-linked traits	0-5	2-20	400
Recessive traits	(very few)	(very small)	1,500

B. Cytogenic (Chromosome Aberration): Effects per million live-born descendants due to 1-rem radiation dose to all individuals in a generation

Effect	Cases in First Generation	Total Cases Over All Future Generations[a]	Current Incidence per Million Live Births
Congenital abnormalities			
Unbalanced rearrangements	12	15	1,000
Aneuploidy	1	1	4,000
Recognized spontaneous abortions			
Aneuploidy and polyploidy	11	11	35,000
XO	3	3	9,000
Unbalanced rearrangements	72	90	11,000

Table 5-12 Continued

C. Total of Effects, Including Genetically Linked Disease (contains values from Parts A and B): Effects per million live-born descendants due to a 1-rem radiation dose to all individuals in a generation

Disease Classification	Cases in First Generation	Total Cases Over All Future Generations[a]	Current Incidence per Million Live Births
Dominant diseases	10-100	50-500	10,000
Chromosomal and recessive diseases	very few	very few	10,000
Congenital anomalies	1-100	10-1,000	15,000
Anomalies expressed later			10,000
Constitutional and degenerative diseases			15,000
Total	11-200	60-1,500	60,000

[a]Assumes population is not increasing or decreasing rapidly so that each generation has the same number of individuals

Public perception of public risk, however, is not completely related to a numerical value. The possible magnitude of the consequence of a single accident is the key factor in the perception of the public.[55-61]

Summary

While the utility industry is confident that its production of electricity by nuclear fission is environmentally sound and safe, this confidence has yet to be transferred to some public sectors. With this public concern in mind, the NRC has placed additional restrictions on nuclear facilities to ensure adequate plant design and proper facility operation.

CONCLUSION

The environmental considerations relative to an energy facility are not black and white. Variability of ecosystems from site to site can cause different evaluations of impact parameters at each site. Although the environment is now a component of the siting process, methods are still being developed to allow a meaningful analysis of design alternatives.

Techniques for relating system design with impact analysis, as illustrated in Figure 5-5, are important to final economic choices. The success of any such system relies on the identification of correct and applicable considerations. The success also depends upon the amount of time and resources available for identification. Effective environmental decisions are determined by the completeness of information gathered, not the amount.

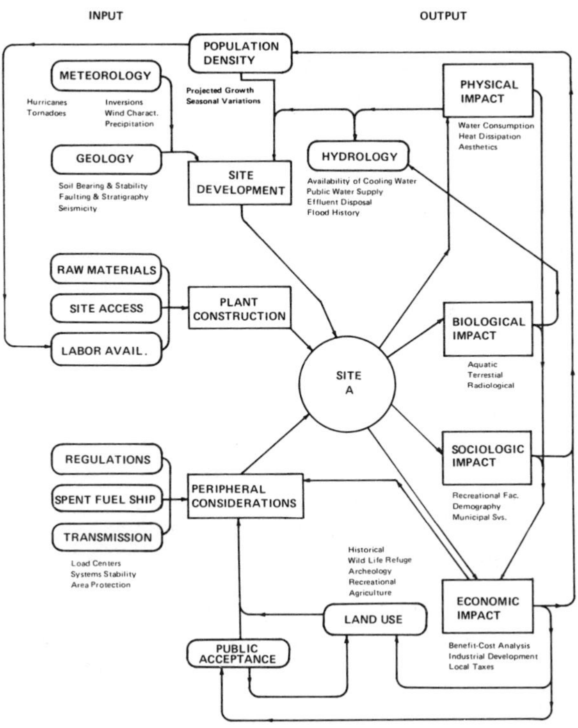

Figure 5-5 Schematic representation of a power plant design process that considers environmental impacts.[62]

REFERENCES

1. "Resource and Land Investigations (RALI) Program: Methodologies for Environmental Analysis, Vol. 2, Power Plant Siting," MITRE Corp. MTR 6740, August 1974.
2. "General Environmental Guidelines for Evaluating and Reporting the Effects of Nuclear Power Plant Site Preparation, Plant and Transmission Facility Construction," Prepared by Hittman Associates, Inc., for the Atomic Industrial Forum, February 1974.
3. "Environmental Effects of Using Coal for Generating Electricity," U.S. Nuclear Regulatory Commission, NUREG-0252, Prepared by Argonne National Laboratory, June 1977.
4. "Power Plant Site Evaluation Interim Report-Douglas Point Site," Vol. 2, Maryland Department of Natural Resources Power Plant Siting Program, Annapolis, Maryland, Prepared by Johns Hopkins University, May 1974.
5. "Improving Regulatory Effectiveness in Federal/State Siting Actions," U.S. Nuclear Regulatory Commission Office of State Programs, NUREG-0195, May 1977.
6. Jensen, L., "The Effects of Thermal Discharges into Surface Water," *Human and Ecologic Effects of Nuclear Power Plants,* Charles C. Thomas, 1974.
7. Rasmussen, M., "Reactor Safety Study: Assessment of Accident Risks in U.S. Commercial Nuclear Power Plants," U.S. Nuclear Regulatory Commission, WASH 1400, 1976.
8. "Nuclear Energy Center Site Survey – 1975, Part III of V, Technical Considerations," U.S. Nuclear Regulatory Commission, NUREG-0001, January 1976.
9. "Preliminary Investigations of a Potential Power Plant Site in Cecil County, Maryland," Vol. 1 – Summary, Maryland Department of Natural Resources Power Plant Siting Program, Annapolis, Maryland, Prepared by Engineering-Science, Inc., February 1974.
10. "Power Plant Site Evaluation, Brandon Shores Site," Maryland Department of Natural Resources Power Plant Siting Program, Annapolis, Maryland, Prepared by Johns Hopkins University, PPSE 1-2, November 1972.
11. "Addendum to PPSE 1-2, Power Plant Site Evaluation, Brandon Shores Site," Maryland Department of Natural Resources Power Plant Siting Program, Annapolis, Maryland, Prepared by Johns Hopkins University, January 18, 1973.

12. "Power Plant Site Evaluation Report, Dickerson Site," Maryland Department of Natural Resources Power Plant Siting Program, Annapolis, Maryland, Prepared by Johns Hopkins University, Addendum to PPSE 3-1, January 1974.

13. "Douglas Point, A Story of Circulation and Mixing," Maryland Department of Natural Resources Power Plant Siting Program, Annapolis, Maryland, Prepared by Chesapeake Bay Institute, May 1975.

14. "Power Plant Site Evaluation Report, Douglas Point Site," Volumes I and II, Maryland Department of Natural Resources, Power Plant Siting Program, Annapolis, Maryland, Prepared by Johns Hopkins University, May 1974.

15. "Power Plant Cumulative Environmental Impact Report," Maryland Depart-of Natural Resources Power Plant Siting Program, PPSP-CEIR-1, September 1975.

16. "Air Quality Criteria for Sulfur Oxides," U.S. Department of Health, Education and Welfare – Public Health Service – National Air Pollution Control Administration, 1969.

17. Hindaul, I., "Air Pollution Injury to Vegetation," U.S. Department of Health, Education and Welfare – Public Health Service – National Air Pollution Control Administration, January 1970.

18. "Air Quality Criteria for Nitrogen Oxides," U.S. Department of Health, Education and Welfare – Public Health Service – National Air Pollution Control Administration, January 1969.

19. "Air Quality Criteria for Particulate Matter," U.S. Department of Health, Education and Welfare – Public Health Service – National Air Pollution Control Administration, January 1969.

20. Norris, J., "Selected Environmental Considerations and their Measuring Parameters," *Proceedings of the Conference on Nuclear Power Plant Siting,* American Nuclear Society, August 26–28, 1974.

21. "Calculated Power Plant Impact on Baltimore Air Shed," Maryland Department of Natural Resources Power Plant Siting Program, Annapolis, Maryland, Prepared by Research Institute for Advanced Studies, January 1973.

22. Dul, E. et al, "Principles of Quantitative Environmental Impact Analysis for Proposed Power Plants," *Proceedings of the International Energy Engineering Congress,* November 1975.

23. "Guidelines for Electric Utilities in Assessing Environmental Effects," Ohio Department of Health, Columbus, Ohio, Prepared by Environmental Research Associates, 1970.

24. "Evaluation of Environmental Effects of a Natural Draft Cooling Tower at the Davis-Bosse Nuclear Power Station, NUS Corporation, July 1971.

25. "Toxicity of Power Plant Chemicals to Aquatic Life," U.S. Atomic Energy Commission, WASH 1249, June 1973.

26. "Analysis and Control of Thermal Pollution," U.S. Environmental Protection Agency, PB 237 617, August 1974.

27. "Development Document for Best Technology Available for the Location, Design, Construction, and Capacity of Cooling Water Intake Structures for Minimizing Adverse Environmental Impact," U.S. Environmental Protection Agency, April 1976.

28. Nelson, G., "Technical and Economic Evaluations of Cooling Systems Blowdown Control Techniques," U.S. Environmental Protection Agency Office of Research and Development, November 1973.

29. "Story of Thermal Discharge Diffusion for the Canadian Great Lakes Shoreline," Canada Department of the Environment Center for Inland Waters, Prepared by Acres Consulting Services, Ltd., March 1974.

30. Jones, N., "Report on the Socio-Economic Impact Issues Working Group," State/Federal Siting Commission Committee Meeting of June 16, 1976.

31. "Socioeconomic Impacts: Nuclear Power Station Siting," U.S. Nuclear Regulatory Commission, NUREG-0150, June 1977.

32. *Proceedings of the Conference on Constructing Social Impact Statements for Power Plant Siting,* Ohio State University, September 1975.

33. Traub, R., "An Investigation on the Socio-Economic Impact of Nuclear Power Plants on Small Communities," Ohio Power Plant Siting Commission, Columbus, Ohio, 1974.

34. "A Survey of Public Leadership Attitudes Toward Nuclear Power Development in the United States," EBASCO Services, Inc., August 1975.

35. "Reactions of the Public and Thought Leaders to a Floating Nuclear Electric Generating Plant," Westinghouse Electric Corp., Prepared by Opinion Research Corp., January 1972.

36. Chang, S. and Snead, R., "An Annotated Bibliography: Social and Economic Factors Associated with Electric Power Generating Stations," Atomic Industrial Forum, October 1976.

37. "State and Local Planning Procedures Dealing with Social and Economics Impacts from Nuclear Power Plants," U.S. Nuclear Regulatory Commission Office of State Programs, NUREG-0203, January 1977.

38. Olsen, M. and Mervin, D., "Toward a Methodology for Constructing Social Impact Statements Using Quality of Social Life Indicators," U.S. Energy Research and Development Administration, Prepared by Battelle Pacific Northwest Laboratory, 1976.

39. Sagan, L. (Ed.), *Human and Ecological Effects of Nuclear Power Plants,* Charles C. Thomas, 1974.

40. "Siting of Nuclear Facilities," International Atomic Energy Agency, Vienna, 1975.

41. Karam, R. and Morgan, K., *Environmental Impact of Nuclear Power Plants,* Pergamon Press, 1976.

42. Ramsey, W. and Reed, P., "Land Use and Nuclear Power Plants Case Studies of Siting Problems," U.S. Atomic Energy Commission Directorate of Regulatory Standards, WASH 1319, October 1974.

43. "State of Ohio Environmental Assessment Program Information Manual," Ohio Department of Natural Resources, Columbus, Ohio, 1972.

44. "Preliminary Criteria for Evaluating Major Utility Facility Applications," Ohio Power Plant Siting Commission, Columbus, Ohio, Prepared by URS Corp., 1975.

45. "Impacts of NEPA and Other Court Actions on the Nuclear Option," *Proceedings of the Atomic Industrial Forum Conference on U.S. Energy Policy,* January 1977.

46. Skipton, E. and Wayland, R., "Comments on 'Economics of Nuclear Power'," Ohio Public Utilities Commission Finance and Economics Section, Columbus, Ohio, December 5, 1975.

47. Weinberg, A., "Energy Policy and Energy Projections: The Case of a Nuclear Moratorium," Institute for Energy Analysis, Oak Ridge Associated Universities, Oak Ridge, Tennessee, January 11, 1977.

48. Chester, C. and Chester, R., "Civil Defense Implications of the U.S. Nuclear Power Industry During a Large Nuclear War in the Year 2000," *Nuclear Technology,* Vol. 31, December 1976.

49. "Regulatory Guide 4.2, Revision 1: Preparation of Environmental Reports for Nuclear Power Stations," U.S. Nuclear Regulatory Commission, January 1975.

50. "An Investigation of Airborne Radioactive Effluent from an Operating Nuclear Fuel Reprocessing Plant," U.S. Department of Health, Education, and Welfare Public Health Service, July 1970.

51. "Observation of Airborne Tritium Waste Discharge from a Nuclear Fuel Reprocessing Plant," U.S. Environmental Protection Agency – Office of Radiation Programs – Field Operations Division, February 1973.

52. "Krypton 85, A Review of the Literature and An Analysis of Radiation Hazards," U.S. Environmental Protection Agency Office of Research and Monitoring, January 1972.

53. "Radioactive Materials Released from Nuclear Power Plants," U.S. Nuclear Regulatory Commission, May 1976.

54. "Report to Congress on Abnormal Occurrences October – December 1975," U.S. Nuclear Regulatory Commission, March 1976.

55. Kendall, H., "Nuclear Power Risks – A Review of the Report of the American Physical Society's Study Group on Light Water Reactor Safety," Union of Concerned Scientists, Cambridge, Massachusetts, June 1975.

56. Kendall, H. and Mowgler, S., "Preliminary Review of the AEC Reactor Safety Study," Joint Review Committee of the Sierra Club (San Francisco, California) and the Union of Concerned Scientists (Cambridge, Massachusetts), November 1974.

57. Dukert, J., "High-level Radioactive Waste: Safe Storage and Ultimate Disposal," U.S. Energy Research and Development Administration, 1975.

58. Hogerton, J., "Nuclear Reactors," U.S. Atomic Energy Commission Division of Technical Information, 1963.

59. Mitchell, W. and Turner, S., "Breeder Reactors," U.S. Atomic Energy Commission Office of Information Services, 76-611325, 1972.

60. "Nuclear Power and the Environment," International Atomic Energy Agency, Vienna, 1973.

61. Coe, R. (Chairman). "Environmental Guidelines," Western Systems Coordinating Council, December 3, 1971.

62. "Electric Power in Texas," Texas Office of the Governor – Division of Planning Coordination – Advisory Committee on Plant Siting, Austin, Texas, November 1972.

6

Future considerations in power plant siting

ENERGY SOURCES

Power plant siting considerations of the future will, to a large degree, be a function of the new energy sources being exploited and the technological modifications introduced to currently used processes. A review of present energy sources and technologies will be presented followed by a review of the possible energy sources and technologies of the future.

Present Sources

Presently, as illustrated in Figure 6-1, electric power is generated by one of two methods: thermal/mechanical generation and mechanical generation. In the former case, fuel is used to produce heat energy in the form of steam to drive a turbine which in turn powers a generator to produce electrical energy. In

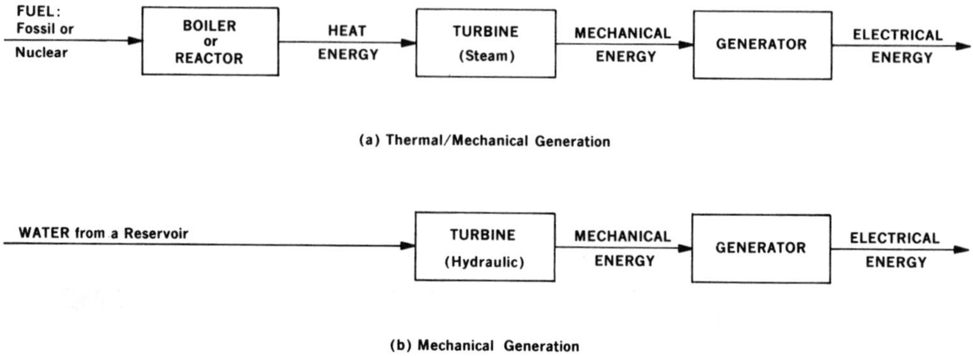

Figure 6-1 Present methods for generating electrical energy.[1]

the latter case, water from a reservoir is used to drive a turbine which in turn powers a generator to produce electrical energy. Electrical energy generation by thermal/mechanical means is inherently inefficient. Efficiencies are typically on the order of 35% to 40%. On the other hand, water power generation of electrical energy has conversion efficiencies in the 80% to 90% range.[2]

Fossil fuel plants represent the most common application of the thermal/mechanical generation concept. Coal, gas, or oil is mixed with air and injected into a boiler where combustion takes place. The heat from combustion is used to convert water to steam. This steam in turn flows through large pipes to a multistage turbine. As the steam passes through the turbine, energy is extracted from the steam to turn the turbine. The turbine, which is mechanically connected to the generator, in turn rotates the generator allowing the production of electrical energy.

Several adjunct stages are present in a modern fossil fuel plant. First, the combustion process produces gases and particles that are emitted to the atmosphere. This emission has been a source of environmental concern and has fostered the consideration of new air emission designs. These designs have included use of taller stacks, catalytic scrubbers, limestone scrubbers, and higher quality fuels. Second, the combustion process produces solid by-products, such as ash, which must be disposed. Third, the turbine stage does not extract all of the heat from the steam. Therefore, the waste heat must be extracted, or the spent steam must be reintroduced to the boiler after passing through a condenser. Where the steam is not reintroduced to the boiler, a cooling system is required.

Nuclear plants also utilize the thermal/mechanical generation concept. The major difference is that a reactor replaces the boiler. Radioactive fuel and control rods are inserted into the reactor. These elements interact to produce heat in a "coolant" which is circulated from the reactor to the turbine to the reactor. Of the many reactor forms that have been developed, three basic types are in commercial use today, (1) the pressurized water reactor (PWR), (2) the boiling water reactor (BWR), and (3) the high temperature gas reactor (HTGR).

While not as prevalent as fossil fuel and nuclear generation, hydroelectric generation plays an important role in electrical energy generation. Hydroelectric generation facilities, in addition to having good efficiencies, are economical to operate, possess a long lifetime, have a rapid response in starting and stopping operations, and require little maintenance. The primary requirement for hydroelectric generation is a water impoundment such as a lake or reservoir. However, with the environmental opposition to initiating new impoundments, this generation mode will decrease in importance in the future.

Future Sources

Energy may be classified in six predominant forms. As noted in Table 6-1, these are chemical, gravitational potential, heat, kinetic, nuclear, and radiation. The energy forms having the highest potential for use in electric power generation are the chemical and nuclear forms. An overview of the sources of all six forms will be presented as background for the second section of this chapter. The section addresses both the federal government's and electric utility industry's policy considerations that will strongly influence the entry of these sources into commercial use in electric energy generation.

Chemical Energy. The sources of chemical energy are numerous. The fossil fuels of coal, gas, and oil have been used in the past. Besides an expanded use of coal in the future, the coal converted products and solid waste could have a position in electric energy generation.

Table 6-1 Potential Sources for Electrical Energy Generation.[1, 2]

Energy Form	Source
1. Chemical	A. Fossil Fuel
	1. Coal
	2. Converted coal
	(a) improved solid
	(b) liquefaction
	(c) low Btu gasification
	(d) high Btu gasification
	3. Crude oil
	4. Natural gas
	5. Oil shale
	6. Tar sands
	B. Fuel cells
	C. Magnetohydrodynamic (MHD)
	D. Solid wastes
	E. Wood
2. Gravitational Potential	A. Hydro reservoir
	B. Tidal
3. Heat	Geothermal
4. Kinetic	Windmills
5. Nuclear	A. Breeder
	B. Fission
	C. Fusion
6. Radiation	Solar

Coal conversion technology is presently being viewed in four categories: improved solids, liquefaction, low Btu gasification, and high Btu gasification. Within each category numerous processes have been devised and are now being researched and developed. The processes generally having the best economics and the earliest point of commercial entry in the electric power industry are those in the improved solids category.

The improvement in solids (coal) is mainly accomplished through beneficiation or solvent refining. In beneficiation, coal is crushed and screened to a maximum size, cleaned to remove noncoal particles and dust, and dried. Both dry and wet washing are used to remove undesirable constituents. As a result, this process produces materials that could present serious disposal problems.

In the solvent refining process, coal is crushed and slurried to a reactor where the coal is dissolved in a hydrogen donor solvent. The solution is then filtered to remove most of the ash. Then it is vacuum flashed to boil off the solvent and solidify the material. In general, combustion of this material will yield fewer adverse environmental impacts than if the coal were combusted directly.

The production of liquid fuels from coal is accomplished by changing the carbon-to-hydrogen ratio (by weight). The major barriers to commercial entry of liquefaction technology will be hardware problems and high economics. Of the process approaches which have been proposed, pyrolysis holds the most promise for use in the electric utility industry. In this case, the char by-product will be used in powerhouse boilers with the liquid product being used as a raw material for the chemical industry.

Gaseous fuels of any energy content (Btu level) can be produced from coal. Generally, the energy ranges of most interest are the low Btu range (100 to 300 Btu) and the high Btu range (approximately 1000 Btus). In the former case, the gas can only be transmitted short distances and must therefore be produced close to the user site. In an electrical energy generation application, the gasification facility must be located close to the power plant, thereby increasing the potential for severe environmental impacts in an immediate geographic region. In the latter case, the gas may be transmitted over long distances. Hence, there is no need for the gasification facility to be near the user.

Low Btu gases are produced in a two-stage process with high Btu gases requiring a third stage. Coal is (1) crushed and screened to reduce the particle size and (2) treated to prevent caking. This composition, along with a hydrogen source and oxygen, is then introduced into a reactor where heat is applied. The coal combusts with the hydrogen and oxygen to form carbon monoxide, methane, and hydrogen. The methane (natural gas) is then extracted. (This methane gas is at the low Btu level. If high Btu gas is required, an upgrading stage is added.) The major barrier to commercial entry of gasification technology will be the economics of the process.

Interest in oil shale as an energy source was kindled in the 1960s and resulted in the U.S. Department of the Interior initiating a

Prototype Oil Shale Leasing Program in the early 1970s. The purpose of the program was to stimulate commercial development of oil shale technology so that the shale's energy content could be extracted as a solid, gas, or liquid fuel. Development of this resource has been hampered and will continue to be hampered until three major barriers are resolved. These barriers are insufficient water, severe adverse environmental impacts, and inefficient conversion of energy.

Tar sands, porous rock containing highly viscous hydrocarbon oils, have also been mentioned as an energy source of the future. The development of tar sand technology lags the technology of other fossil fuel sources. Tar sand mining is especially deficient. In addition, development of this resource will be hampered because of its severe adverse environmental impacts.

Nonfossil fuel chemical energy forms include fuel cells, magnetohydrodymanics, solid wastes, and wood. The fuel cell produces direct current (dc) electricity directly by electrochemically oxidizing hydrogen. The fuel cell is attractive because of its high theoretical efficiency. While this technology has been demonstrated, its potential for utility application in this century is low because of economic and environmental barriers. The requirements for a hydrogen source, continuous operation, and conversion to alternative current (ac) raise questions as to whether this approach holds any advantage over presently used power plant technology.

Magnetohydrodynamics (MHD), while being theoretically attractive, is some years away from being realized commercially. This process uses an ionized gas moving through a magnetic field to produce electrical energy in the direct current (dc) form. The exhaust gases from the conversion process could be used to heat "outside" air which could be processed through a steam turbine to generate electricity.

MHD could improve its commercial viability through a marriage with coal pyrolysis technology. The char by-product of pyrolysis could be used as the main feedstock for MHD with the liquid and gaseous products being used as energy sources for various MHD processing stages. Favorable energy balance situations on the commercial scale could greatly speed up the commercial entry of MHD.

Of the nonfossil fuel chemical energy sources, solid waste possesses the greatest near-term potential for impacting the electric utility industry. Already, major experiments are underway to utilize this source of energy. Most of these experiments, however, use solid waste as an additive in coal fired plants. While the use of solid waste assists in the alleviation of the disposal problem associated with that societal by-product, the economics of its use are a barrier to large scale use. In particular, collection, handling, and classification of the waste is expensive. An additional problem relates to the lack of consistency of the Btu content of the waste.

Gravitational Energy. Energy in the form of gravitational potential is available from two sources. One source, the use of hydro reservoirs in the generation of electricity was discussed earlier in this chapter. While there is little potential for new water impoundments in the form of lakes, the use of reservoirs with pumped storage facilities does hold some promise.

The goal of a pumped storage facility is to pump water from a low elevation to a high elevation during periods of low electrical demand and then to use the gravitational potential of this stored water to produce electricity during periods of high electrical demand. This results in a partial elimination of the peaks and valleys present during a normal day's operation and of the construction of expensive fossil-fuel peaking units. The primary barrier to this energy source will be resistance from environmental groups to the siting of the reservoirs.

For centuries tidal power has been viewed as a source of energy. Unfortunately, tidal power has been plagued with geographic limitations, technological limitations, and the potential of costly capital investments. Unless these issues can be resolved, tidal power will continue to be given little consideration as a source for electrical energy generation.

Heat Energy. Heat energy within the earth's interior forms the basis for geothermal energy. As heat from the earth's magma (molten rock) moves into the earth's lower layers of porous rock, the rock acts as a boiler converting water into hot water, wet steam, or dry steam. These substances sometimes rise to the earth's surface naturally to form geysers. Most of the time, however, these substances remain trapped in the ground in aquifers where removal can be accomplished only by drilling.

While this potential source appears to be a most attractive heat energy source, certain technical problems presently prevent acceptance of this energy for use in electric power generation. First, mineral concentration in the geothermal fluid is high. These impurities tend to react unfavorably with transport system components thereby creating significant hardware problems. These problems must be solved before wide-scale application occurs. Second, much of the steam possesses temperatures lower than the temperatures currently used by the utility industry. As a result, modifications will be required in turbine designs. Third, the environmental problems related to this energy source are of great concern to many. Of particular concern are the disposal of the spent fluid and the problem of subsidence (the sinking of the earth's crust above the geothermal aquifer).

An interesting sidelight relates to the potential of geothermal fluids as a source of energy in other than the heat form. Specifically, in the Gulf Coast region aquifers are present that contain fluids composed of steam, natural gas, and oil. This fluid could prove to be a most valuable source of natural gas.[3]

Kinetic Energy. Kinetic energy in the form of wind is present at varying times all across the earth's surface. Unfortunately, it lacks the potential for impacting the electric utility industry. Regions of consistent, high velocity wind are scarce and, as a general rule, remote from electrical load centers. In particular, this source does not satisfy power system reliability requirements.

Nuclear Energy. Nuclear energy is currently viewed from the perspective of its three sources. (1) *Nuclear fission,* the process whereby energy is released when a heavy atom is split into two dissimilar atoms, provides the basis for modern day nuclear power plants. (2) *Nuclear breeder process,* a process whereby Uranium 238 is converted to Plutonium, provides the basis for a future power source that meets its own fuel requirements in the generation process. (3) *Nuclear fusion,* a process that releases energy when two light nuclei are joined, provides the potential for a major power source in the next century. While the nuclear fission facilities of today produce many environmental and safety related concerns in the minds of the public, the breeder and fusion concepts of the future should tend to alleviate some of the concern. For instance, the breeder will not require the quantities of newly mined uranium that are currently required by the fission process. On the other hand, fusion facilities, if current theory and laboratory work can be extrapolated to the commercial level, will mitigate many of the environmental concerns existing with fission facilities.[4]

The future of nuclear energy is not bright. Recent federal policy shifts have curtailed development of new technologies. Increased intervention by environmentalists in the power plant siting phase and increased difficulties with the nuclear fuel supply system have threatened the near-in position of nuclear fission.[5] Nuclear fuel supply is a special concern.

As presented in Figure 6-2, nuclear fuel is processed in ten steps. After uranium ore is mined, it is concentrated in a milling operation and then refined into a form known as "yellow cake." At this point the fuel enters what may be referred to as the "nuclear fuel cycle."

The first step of the fuel cycle is enrichment, the concentration of Uranium 235 by separating this isotope from Uranium 238. The concentrated Uranium 235 is then converted to uranium dioxide powder and fabricated into rods. These fabricated rods are placed in a nuclear reactor to serve as a portion of the heat source in electric-power generation. After approximately four years, these rods are removed and placed in storage until reprocessing can take place. The

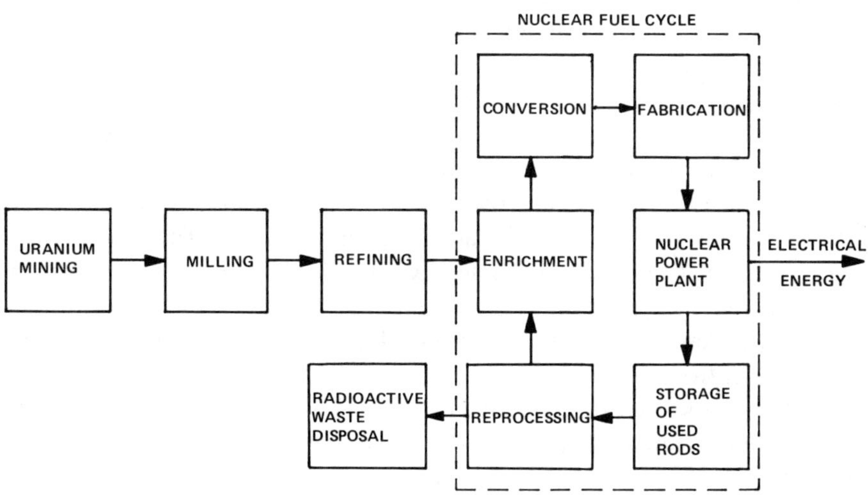

Figure 6-2 Nuclear fuel supply system.

reprocessing step allows much of the unused uranium to be reclaimed for reinsertion into the enrichment step. Radioactive waste by-products of the reprocessing step are disposed of in approved areas.

While the nuclear fuel supply system of Figure 6-2 appears to be sound, in actuality many problems presently exist that, if not solved, will impair the position of the fission process in future electrical energy generation. First, the cost of the raw uranium fuel has increased appreciably in five years. Some suppliers have increased costs by as much as 700%. Second, the enrichment capacity within the United States is not of sufficient size to handle the requirements of both the military and the private sectors. Third, there is currently no domestic reprocessing facility available, and none is realistically projected until the mid 1990s. As a result, all used fuel rods are being retained by the electrical utility industry in storage facilities with no rods being reprocessed. Fourth, with the lack of reprocessing many utilities are reaching their capacity to store used rods. Finally, radioactive waste disposal procedures and sites are being questioned. Each of these problem areas have become sufficiently critical to threaten the closing down of current operational facilities and the cancellation of planned facilities.

Radiation Energy. Solar energy has frequently been discussed as an alternative to present energy sources. While solar radiation from the sun is continuous, its level of intensity on the earth is not constant. Variations occur due to atmospheric conditions, the time of the day, and the season of the year. A second complication occurs with respect to collection systems and storage systems. Radiation energy collection systems are expensive and bulky. At the energy levels required for electric power system applications, solar systems would be cost prohibitive. In addition, mechanisms for solar energy storage have not been sufficiently developed for power system applications.

The apparent long-term position of solar energy appears to rest with applications by individual consumers in the residential and commercial sectors. Units appropriately designed to meet the needs of the user should provide a viable economic alternative to other energy sources.

Any reader desiring additional background material relative to energy sources should consult References 1 and 2.

POLICY CONSIDERATIONS

Development of new energy sources and energy related technology will be a function of the energy research and development policies of the federal government and the electric power industry. The research and development policies of the federal government are prepared and propagated annually by the U.S. Energy Research and Development Administration (ERDA) in a document called the National Plan for Energy Research, Development, and Demonstration.[6] The research and development policies of the electric power industry are prepared and propagated by the Electric Power Research Institute (EPRI), an organization supported by more than 500 member organizations.[7] The policies developed by these agencies will now be considered.

Federal Government View

The policy of the federal government is based on two principles. First, since the private sector produces and consumes most of the energy, it is felt that this sector has the primary responsibility to solve energy related problems. Second, it is felt that the nation's market forces can provide the most effective means for resolving the energy crisis. Federal programs, therefore, are intended to complement the technological activities of the private sector.

While the federal government allocates responsibility for developing commercial technologies to the private sector, it gives guidance to this development by providing funds for processes and technologies that seem to hold the greatest promise. For instance, since there is an existing national goal to reduce dependence on gas and oil, six key technologies have been identified and are being funded that will increase energy supply. These technologies are (1) direct utilization of coal, (2) synthetic fuels from coal, (3) enhanced gas and oil recovery, (4) nuclear light water reactors, (5) geothermal energy, and (6) solar heating and cooling.

Federal policy is developed according to specific time periods. These time periods are called near-term (now to 1985), mid-term (1985–2000), and long-term (beyond 2000). The near-term policy is oriented toward increasing the use of coal and nuclear energy, improving coal combustion technology, solving many of the problems related to nuclear fuel supply referenced in the previous section of this chapter, and enhancing gas and oil recovery. The mid-term policy highlights the production of synthetic fuels from coal, extraction of oil from shale, geothermal energy utilization, and solar heating and cooling. The long-term policy envisions development of more exotic technologies such as the breeder reactor, hydrogen fusion, and solar electricity. While federal policy tends to be broad in scope and lacking in definition, the policy of the electric utility industry has been developed to provide the specific definition required by the industry to fulfill its role in the federal program.

Utility Industry Thrust

The near-term, mid-term, and long-term goals of EPRI parallel the technology thrusts of the federal government. The major difference is that EPRI's goals have been directed to the electric utility power plant setting. Research areas have been specified to meet these goals, and programs have been developed in each research area. Tables 6-2 and 6-3 outline the program thrusts of these research areas for the near-term and the mid-term. A review of these tables will give the reader insight into the technologies the electric utility industry is relying upon for its production of electric power through the end of this century.

Table 6-2 EPRI Program Thrusts for the Near-term.[7]

Research Area	Program Thrusts
A. Primary energy resource processing	1. Gasification of coal 2. Liquefaction of coal 3. Direct utilization of coal 4. Environmental control and combustion of coal processes
B. Energy conversion	1. Nuclear water reactor technology 2. Nuclear reliability, availability, and economics 3. Nuclear fuels, waste, and environmental effects

	4. Thermal-mechanical energy conversion and storage 5. Hardware development through the electric utility equipment spectrum
C. Electrical systems	1. Electric power transmission: ac overhead, dc, and underground 2. Distribution 3. System planning, security, and control 4. Rotating electrical machinery
D. Environment and conservation	1. Heat pump energy conversion and storage 2. Solar energy heating and cooling 3. Energy demand and conservation 4. Electric power transmission: ac overhead and underground 5. Environmental control and combustion 6. Environmental assessment

Table 6-3 EPRI Program Thrusts for the Mid-term.[7]

Research Area	Program Thrusts
A. Primary energy resource processing	1. Gasification of coal 2. Liquefaction of coal 3. Direct utilization of coal 4. Environmental control and combustion of coal 5. Nuclear fuels, waste, and environment 6. Developing nuclear applications and technology
B. Energy conversion	1. Electrochemical energy conversion and storage 2. Thermal-mechanical energy conversion and storage 3. Geothermal energy 4. Developing nuclear applications and technology 5. Environmental assessment

Table 6-3 Continued.

	6. Water reactor system technology 7. Environmental control and combustion
C. Electrical systems	1. Electric power transmission: ac overhead, dc, and underground 2. Distribution 3. System planning, security, and control 4. Rotating electrical machinery
D. Environmental and conservation	1. Electric power transmission: ac overhead and underground 2. Environmental control and combustion 3. Environmental assessment

DESIGN ALTERNATIVES

As displayed in Figure 6-1, present electrical energy generation is based primarily upon (1) the use of fossil and nuclear fuels to generate heat to produce the mechanical energy that in turn produces electricity and (2) the use of water to drive a turbine to produce the mechanical energy that in turn produces electricity. Many of the technologies now being considered are, in essence, variations to conventional fossil fuel boiler-fired power plants.

Figure 6-3 presents the major components of conventional fossil fuel boiler-fired power plants. Fossil fuel in solid, liquid, or gaseous form is fed to a boiler where the fuel is burned and steam is produced. The by-products of combustion are vented to the atmosphere after being "cleaned." The steam is used to drive a turbine to produce mechanical energy. As the energy in the steam is expended and the steam loses its usefulness in the turbine, it is cooled and returned to the water state. The mechanical energy from the turbine is used to drive the generator to produce electricity.

Alternatives for Fossil Fuel Boiler-Fired Plants

A number of the proposed alternatives for fossil fuel plants will influence future power plant siting considerations. These alternatives

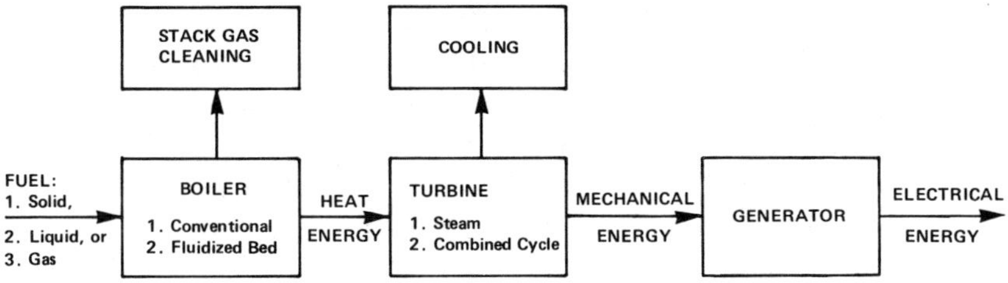

Figure 6-3 Alternatives for a conventional fossil fuel boiler-fired plant.[1]

include fluidized bed boilers, the various stack gas cleaning technologies, binary cycle turbines, the various cooling technologies, and new heat energy sources.

Boiler Alternatives. In conventional boilers the fuel is blown into a furnace through burners which sustain an intense flame within the boiler. As the fuel is combusted, sulfur oxides (SO_x), nitrogen oxides (NO_x), and particulates are produced. If not controlled, the by-products are emitted to the atmosphere. In addition, an ash and slag are produced that require disposal. The boiler design and the type of fuel used are the primary factors determining the amount and character of these by-products.

While fluidized bed boilers are not presently available commercially, the technology is touted as being able to alleviate many of the environmental problems surrounding present boiler designs. The fluidized bed boiler passes its fuel (usually coal), along with a noncombustible material, over a grate which is subjected to an upward air movement. Thus fuel–air mixture is combusted to heat water to produce steam. This boiler design is claimed (1) to be capable of having higher efficiencies than conventional boiler designs, (2) to have the ability to burn high-sulfur coals, producing by-products that are less adverse environmentally than conventional boilers, and (3) to be more economical to construct than conventional boilers.

Alternate Stack Gas Cleaning Technologies. While stack gas cleaning has received appreciable consideration as a means for improving the quality of the gaseous effluent from boilers and while some are available commercially, it is still a developing technology. The technology primarily centers upon the removal of three combustion by-products: sulfur oxides, nitrogen oxides, and particulates.

Sulfur oxides can be removed by a variety of processes. The most frequently referenced technology is the "scrubber." This process passes the stack gas through a material, most often lime or limestone, to react the sulfur dioxide to produce a "solid" compound that can be disposed. While proponents emphasize the ability of scrubbers to improve the quality of the air effluent, opponents question the economics of these processes, process efficiencies, and the potential adverse environmental impact of the waste by-product from these systems.

Removal of nitrogen oxides are primarily affected by boiler design and operating conditions. While catalytic scrubbers are proposed for nitrogen oxide stack gas cleaning, these scrubbers are expensive.

Particulate removal is usually accomplished mechanically or electrostatically. In mechanical separation the gas effluent is accelerated so that the particles can be "slung" out of the gas composition and collected by water or in a hopper. In electrostatic separation the particles are ionized and subjected to a strong electric field where they are attracted to a collection plate. The particles are then cleaned from the plates by a rapping action. Particulate removal systems vary widely in efficiencies.

Turbine Alternatives. In the electric power generation process the steam turbine is the most inefficient component. Several alternative designs have been proposed and hold promise for improving turbine efficiency. These approaches include the use of heat engine cycles covering varying parts of the temperature range, topping cycles, and bottoming cycles.

One approach currently being pursued is the "combined cycle" approach. The combined cycle concept combines a gas turbine with a steam turbine. The exhaust from the gas turbine is used to produce steam. This steam is combined with steam produced by a conventional heater to drive a steam turbine. This approach is believed to be capable of achieving efficiencies higher than obtained currently by conventional turbines.

Cooling Alternatives. An interesting area of concern in power plant designs relates to the dissipation of waste heat from turbines. Debate revolves about the amount of heat and water that should be used in the dissipation process, and the type of process that should be used. Presently four systems are utilized: cooling ponds, once-through cooling systems, wet towers, and dry towers.

Cooling ponds provide a holding area for turbine steam which has condensed to water. The pond is designed so the spent turbine water and water from an external source will circulate through the pond, allowing cooling by evaporation, to a point where the water can be reused. Once-through cooling systems, on the other hand, are designed with the intent that water is never recycled to the plant. While this latter approach is most economical, water requirements are significantly higher than with a cooling pond.

Cooling towers present an expensive alternative when water supply is scarce or when there is insufficient land area for a cooling pond. The towers are designed so that waste heat is dissipated from the spent water by evaporation as air is brought in contact with the water. The air and water come in direct contact in wet towers. In dry towers there is no direct contact.

Alternative Heat Energy Source. Design alternatives being examined for future use also include alternate energy sources. Included in this consideration are varying forms of nuclear reactors, geothermal energy, and solar energy as displayed in Figure 6-4. The viability of these alternatives to fossil fuel was discussed earlier in this chapter.

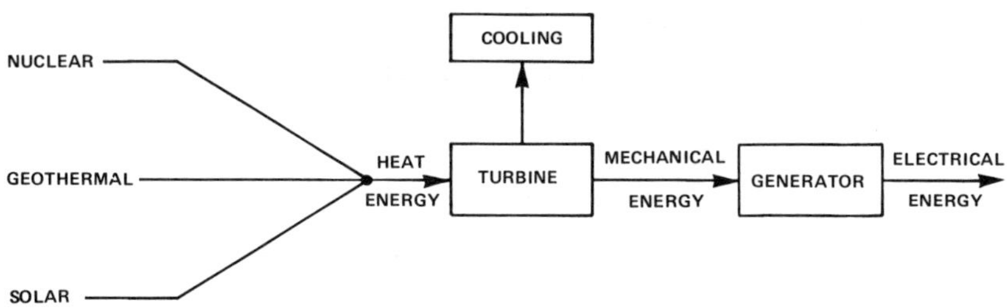

Figure 6-4 Alternative sources of heat energy for boiler-fired power plants.[1]

Alternative Systems for Electrical Generation from Energy in the Chemical Form

Figure 6-5 illustrates the alternatives available for future electrical energy when energy in the chemical form is considered as the source. Each of the four technological alternatives to boiler-fired plants is presently being investigated with the hope that commercial application will someday be achieved or expanded. The positions of two of the alternatives, fuel cell technology and magnetohydrodynamics technology, were discussed earlier in this chapter.

Gas turbine technology has already achieved commercial entry on a small scale. Gas turbines utilize compressed air and either a gas or a vaporized liquid in a combustion chamber to produce a high pressure

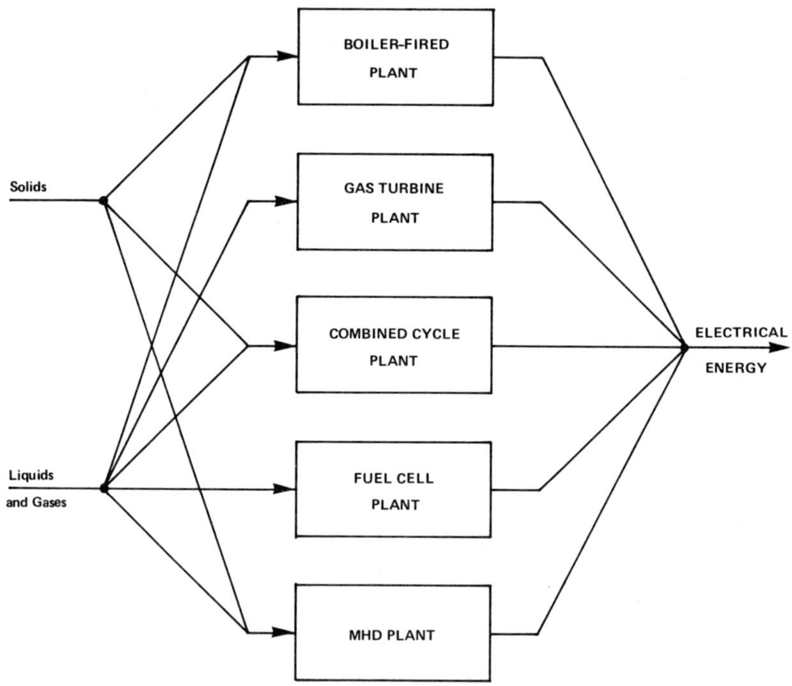

Figure 6-5 Alternatives for electrical generation using energy in the chemical form.[1]

gas that will drive a turbine. Since exhaust gases can be vented directly to the atmosphere, no cooling is required. The process, however, requires a clean burning fuel so that the turbine components will not be appreciably contaminated. The use of a clean burning fuel does offer the advantage that environmental contaminants are reduced. Low efficiency and the high cost of clean fuels are a serious disadvantage for this technology.

The combined technology was referenced earlier in this chapter. Combined cycle systems exist both in boiler-fired processes and in unfired boiler processes. The approach discussed earlier related to the boiler-fired process. The form of combined cycle technology of greatest current interest involves the unfired boiler approach. As in the fired boiler approach, both gas and steam turbines are utilized. The hot exhaust from the gas turbine is used to generate steam in an unfired boiler. The steam then drives a conventional turbine. This approach is believed to be capable of achieving efficiencies in the 40% range.

SYSTEMS VIEW OF POWER PLANT SITING

Rationale for Systems View

With the emphasis toward new energy sources, more efficient processes, and stricter environmental standards, power plant siting will require an increasing use of the "systems approach" in evaluating a site and its alternatives. Of the numerous approaches proposed, the MERES (Matrix of Environmental Residuals for Energy Systems) system is an excellent example of a system approach to energy system evaluation.[8]

As outlined in the MERES system, the production of energy for use by a consumer requires numerous handling and processing steps. For example, as noted in Figure 6-6, the flow of energy from resource form to the consumer involves resource extraction, several processing steps, various transportation steps, and storage at several different locations. Evaluation of a power plant site separate from its place in the entire energy production setting could lead to serious problems in future years.

The example system presented in Figure 6-6 illustrates this last point. The location of the gas-fired electrical generation facility at a

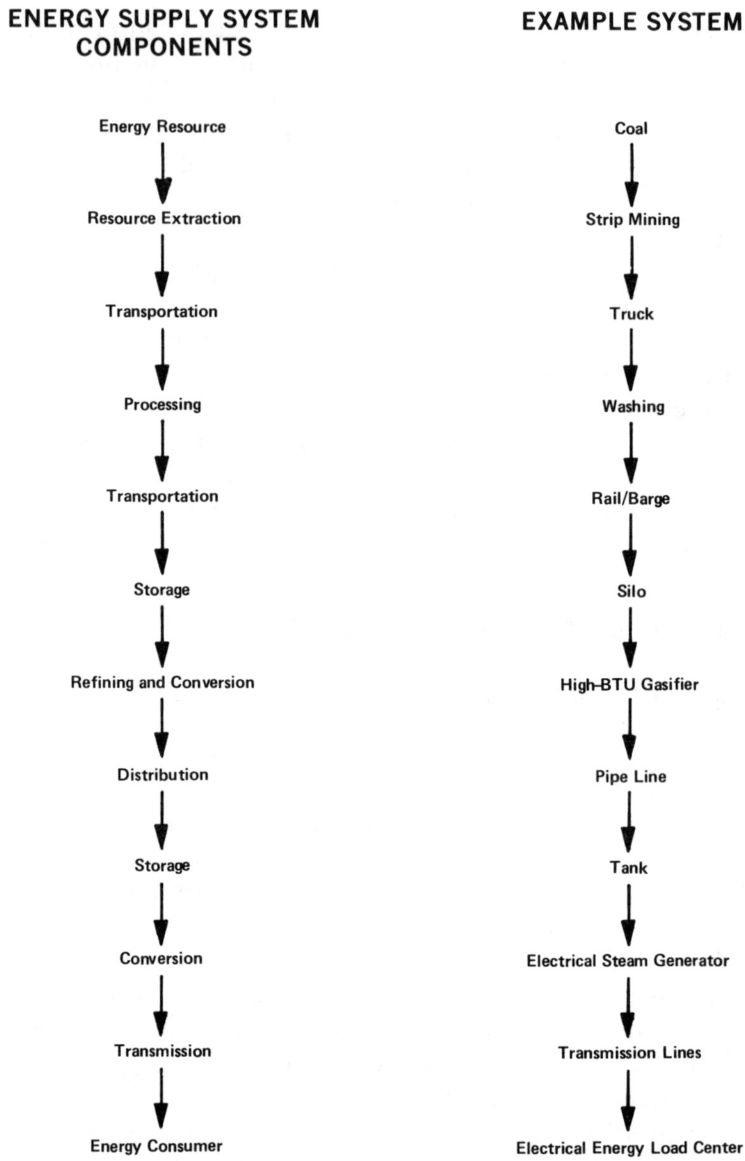

Figure 6-6 Components of an energy supply system and example.[8]

particular site could, on the whole, have a more adverse effect on the environment than, say, a coal-fired plant. This effect could be overlooked if the siting evaluation were limited to the generating plant as opposed to analysis of the entire energy supply system. In the case sited, the aggregate environmental degradation and the aggregate economics from both the high-Btu gasification facility and the power plant could be worse than the environmental degradation and the economics for a coal-fired plant. Individual consideration of the high-Btu gasification facility and the gas-fired electrical plant probably would not have revealed this fact.

This illustration points to a most important need regarding future power plant siting evaluations. The legislative authorization for power plant siting evaluation (1) must encompass all forms of power plants (not just be limited to plants which convert an energy resource to electricity) and (2) must insure that the evaluator has the ability (a) to evaluate the entire energy supply system supporting the plant being reviewed and (b) to factor the results of this review into the evaluation process.

Procedure for Comparing Alternative Energy Systems

Comparison of alternative energy supply systems should not begin until two areas are properly defined. The first pertains to the objectives of the energy supply system. These objectives should contain a statement of the proposed facility's desired traits and the constraints to be imposed on all of the system components. The second area relates to the specification of the particular concerns that are to be evaluated. For example, in evaluating air pollution, points of comparison could be sulfur oxides, nitrogen oxides, particulates, carbon monoxide, carbon dioxide, aldehydes, and hydrocarbons. Each area of environmental and economic concern that is to be reviewed should be listed at the beginning of the evaluation.

After stating objectives and selecting points of comparison, the alternatives to be compared should be selected. The number of possible alternatives available will far exceed the number that can logically be analyzed. Unfortunately, in recent years the requirement for evaluation of alternatives has proliferated beyond the manageable level. If this trend is not curbed, the alternatives of future years could provide so many new variations to the evaluation procedure that analysis and evaluation would be completely impossible to accomplish.

A "system approach" could alleviate this stumbling block. Specifically, of all the possible permutations available for the energy system housing a specific power plant, only the three most viable variations should be evaluated. These viable alternatives should be sufficiently diversified that major trade-offs could be compared. The analysis should then produce a recommended energy system that would provide the setting for the power plant under construction.

Next, the analysis should focus on the technological alternatives within the recommended energy system. This analysis should result in a specific definition of each technology to be used so that the impact of the integrated system can be developed. The basis for selection for each technology should be traceable to specific economic or environmental benefits. In essence, the system, as defined, should approach the optimum design mandated by the system objectives. At this point a clear and meaningful decision should be made concerning the acceptability of the proposed power facility.

Presentation of Results

As a final point, the importance of the proper presentation of evaluation data and results should be noted. The key to an informative evaluation rests with the form in which the data and results are displayed. Both graphical and numerical comparison techniques have been developed. Graphical comparison techniques include computer map analysis, map overlays, shaded matrices, value profiles, relative merit analyses, physical models, and contour mapping. Numerical methods include ranking systems, expected value methods, value matrices, rating systems, utility theory applications, and decision models.[9]

Much of the confusion generated by comparative studies is the result of the absence of a clearly defined method for presenting evaluation data and results. A clearly defined industry or governmental standard is needed in this area. At the very minimum, the evaluation agency having the lead responsibility within each state should take the initiative to develop such guidelines.

CONCLUDING THOUGHTS

Throughout this book, an attempt has been made to establish the basic criteria and regulations that affect the siting of electric generating facilities. Many of the constraints that are placed upon both the process and the public have been discussed. In addition, an attempt has been made to identify some of the conflicting ideologies and objectives. Finally, this chapter has focused on the technological alternatives that are available for the future and how these alternatives are being viewed.

Power plant siting, or the siting of any large energy generation source, is a dynamic process. Changes in emphasis and scope are affected by public opinion, governmental regulation, and technological advances. Note the evolution in federal environmental legislation as it relates to power plant siting. From an initial emphasis on air quality, emphasis moved to water pollution control and then to solid waste regulation. A similar evolution has occurred in governmental regulation of electric utilities. From its initial antitrust emphasis, the federal government moved to regulation of utility operation and financing. Now states are also involved in utility regulation in the areas of financing, construction, operation, and management procedures.[10] In the area of technological change, power plant types have advanced from hydroelectric facilities to fossil-fuel facilities to nuclear facilities.

Power plant siting presently faces a new emphasis. Over the past two decades, technological changes have advanced the rate of energy consumption and production. This energy growth has been directly coupled to economic growth.[11] The conservation efforts of 1974 and 1975, however, altered this pattern by reducing the energy load growth for that period. Now the need for further energy expansion is being questioned. The effects of energy conservation are now being introduced into all energy considerations.

At the same time the effects of conservation on energy demand are being considered, two factors must be kept in sharp focus. First, while the rate of energy growth will probably be much slower in the years ahead, energy growth will, nevertheless, still occur.[12-14] This energy growth will be especially felt in the area of electrical energy usage with new industries and residences using electricity as opposed to natural gas and oil. Second, many old electrical generating facilities will become technically obsolete and uneconomical to operate. These facilities will have to be replaced. Hence, there will be a definite, continuing need for new electrical generating facilities.

In order to deal more effectively with the siting of these new facilities, new alternative energy sources must be considered,[10, 15] present generation concepts must be questioned,[16] the methods used in the siting process must be analyzed[17-19] and optimized,[20] cooperation must be encouraged between electric utilities, [21-23] and structural reform of the electric utility industry must be considered.[24] It is hoped that this book will aid individuals interested in the siting of new power plants to understand the present state-of-the-art of power plant siting and will orient these individuals toward the mechanism through which involvement can occur.

REFERENCES

1. *Energy Alternatives: A Comparative Analysis,* U.S. Government Printing Office (Stock No. 041-011-00025-4, Catalog No. PRES 14.2:EN2), May 1975.
2. Healy, T., *Energy, Electric Power, and Man,* Boyd and Fraser, 1974.
3. "Geothermal Energy Promising But Risky," *Chemical and Engineering News,* Vol. 54, No. 42, October 11, 1976.
4. "Fusion Focus Shifts to Practical Aspects," *Chemical and Engineering News,* Vol. 54, No. 42, October 11, 1976.
5. "The Importance of Closing the Nuclear Fuel Cycle," Atomic Industrial Forum, October 1, 1976.
6. "Creating Energy Choices for the Future," U.S. Energy Research and Development Administration, 1976.
7. "A Summary of Program Emphasis for 1976," Electric Power Research Institute.
8. "MERES and the Evaluation of Energy Alternatives," U.S. Government Printing Office (Stock No. 041-011-00026-2, Catalog No. PREX 14.2:M42).
9. Hagerty, J. and Conner, D., "Environmental Considerations in Power Plant Siting," *Proceedings of the Second Iranian Congress of Civil Engineering,* Pahlavi University, May 3-7, 1976.

10. "Geothermal Energy Program," U.S. National Science Foundation, March 1976.

11. Schurr, S., *Energy, Economic Growth and the Environment,* Johns Hopkins University Press, 1971.

12. "Energy Conservation's Impact on R & D," *Business Week,* June 27, 1977.

13. "Will Energy Conservation Throttle Economic Growth?," *Business Week,* April 25, 1977.

14. Alexander, T., "Industry Can Save Energy Without Stunting Its Growth," *Fortune,* May 1977.

15. "The Government Gambles on the Centrifuge," *Business Week,* May 30, 1977.

16. "U.S. Nuclear Power May Be on the Wrong Track," *Business Week,* April 18, 1977.

17. "Early Site Reviews for Nuclear Power Facilities," U.S. Nuclear Regulatory Commission Office of Nuclear Reactor Regulation, NUREG-0180, May 1977.

18. Doub, W. (Chairman), "Federal Energy Regulation – An Organizational Study," Federal Energy Regulation Study Team, U.S. Governmental Printing Office Document Number 5210-00386, April 1974.

19. "Energy and the Law: Problems and Challenges of the Late 70's," American Law Institute – American Bar Association Study Course, 1977.

20. "Improving Regulatory Effectiveness in Federal/State Siting Actions," U.S. Nuclear Regulatory Commission Office of State Programs, NUREG-0195, May 1977.

21. "6th Annual Review of Overall Reliability and Adequacy of the North American Bulk Power System," National Electric Reliability Council, 1976.

22. "Fossil and Nuclear Fuel for the Electric Utility, Generation Requirements and Constraints 1976-1985," National Electric Reliability Council, 1976.

23. "Factors Affecting the Electric Power Supply 1980-85," U.S. Federal Power Commission Bureau of Power, 1976.

24. "Structural Reform in the Electric Power Industry," U.S. Federal Energy Administration, G-76/323, 1976.

Appendix I

ENERGY BASICS*

A. Conversions of Energy Units

1 British thermal unit (Btu)	=	.000393 horsepower-hours (hph) .000293 kilowatt-hours (kWh)
1 horsepower-hour (hph)	=	2545 British thermal units (Btu) .746 kilowatt-hours (kWh)
1 kilowatt-hour (kWh)	=	3410 Britith thermal units (Btu) 1.34 horsepower-hours (hph)

B. *Approximate* Btu Values of Standard Energy Sources

1 barrel of crude oil	= 5,800,000 Btu
1 standard cubic foot of natural gas (pipeline gas)	= 1030 Btu
1 ton of bituminous coal (Kentucky)	= 25,000,000 Btu
1 pound of bituminous coal (Kentucky)	= 10,000 to 14,000 Btu

C. *Approximate* Btu Ranges Associated with Standard Energy Terms

Low Btu gas – 100 to 300 Btu/cu. ft.[†‡]

Medium Btu gas – 500 to 700 Btu/cu. ft.[‡]

High Btu gas (pipeline gas) – greater than 1000 Btu/cu. ft.

Synthene (pipeline gas made synthetically) – greater than 1000 Btu/cu. ft.

Town gas (the same as low Btu gas) – 200 to 300 Btu/cu. ft.

D. *Approximate* Conversions for Bituminous Coal (Kentucky)

= 12,000 to 15,000 cu. ft. of Synthene

25,000 to 30,000 cu. ft. of medium Btu gas

70,000 to 130,000 cu. ft. of low Btu gas

3 to 3½ barrels of crude oil or liquid

*A portion of these conversions were supplied by Eugene V. Plock.

†In the USSR low Btu gas may run as low as 20 to 50 Btu/cu. ft.

‡These terms are in the development phase. Hence, Btu definitions may change with time.

Appendix II

REFERENCE SOURCES

The following state agencies may be contacted for additional reference material on power plant siting.

ARIZONA	Office of Economic Planning and Development 1624 West Adams Phoenix, AZ 85007
ARKANSAS	Public Service Commission Justice Building Little Rock, AR 72201
CALIFORNIA	Facilities Siting Division California Energy Resources, Conservation, and Development Commission 111 Howe Avenue Sacramento, CA 95825

CONNECTICUT	Power Facility Evaluation Council State Office Building Hartford, CT 06115
FLORIDA	Power Plant Siting Program Department of Environmental Regulation 725 South Bronough Tallahassee, FL 32301
ILLINOIS	Illinois Commerce Commission 527 East Capital Avenue Springfield, IL 62706
IOWA	Iowa Commerce Commission 300 Fourth Street Des Moines, IA 50309
KANSAS	Kansas Corporation Commission State Office Building Topeka, KS 66603
KENTUCKY	Public Service Commission Capital Plaza Tower Frankfort, KY 40601
MAINE	Department of Environmental Protection Bureau of Land Quality Control State House Augusta, ME 04333
MARYLAND	Power Plant Siting Program Department of Natural Resources State Office Building Annapolis, MD 21401
MASSACHUSETTS	Energy Facilities Siting Council John W. McCormack Building One Ashburton Place Boston, MA 02108

MINNESOTA	Environmental Quality Council Capital Square Building 550 Cedar Street St. Paul, MN 55101
MONTANA	Department of Natural Resources and Conservation Energy Planning Division 32 South Ewing Natural Resources Building Helena, MT 59601
NEBRASKA	Power Review Board Department of Water Resources Room 305 1220 J Street P.O. Box 94721 Lincoln, NE 68509
NEVADA	Public Service Commission 505 East King Carson City, NV 89710
NEW HAMPSHIRE	Site Evaluation Committee Prescott Park 105 Loudon Road Concord, NH 03301
NEW MEXICO	Public Service Commission Bataan Memorial Building Santa Fe, NM 87503
NEW YORK	State Board of Electric Conservation Public Service Commission 50 Wolf Road Albany, NY 12205
NORTH DAKOTA	Siting and Reclamation Division State Capital Building Bismarck, ND 48505

OHIO	Power Siting Commission 361 East Broad Columbus, OH 43216
OREGON	Department of Energy 4263 Commercial, S.E. Salem, OR 97310
SOUTH CAROLINA	Public Service Commission P.O. Box 11649 Columbia, SC 29211
VERMONT	Public Service Board 7 School Street Montpelier, VT 05602
VIRGINIA	Division of Public Utilities State Corporation Commission Box 1197 Richmond, VA 23209
WASHINGTON	Energy Facility Site Evaluation Council 820 East 5th Avenue Olympia, WA 98501
WISCONSIN	Bureau of Environmental Energy Systems Public Service Commission Hill Farms State Office Madison, WI 53702
WYOMING	Office of Industrial Siting Suite 500 Boyd Building Cheyenne, WY 82002

Appendix III

THE GENERATION OF ELECTRICITY

The concept of generating electric power by the use of steam has existed for a number of years. While the fundamentals are extremely simple, the actual physical plant is quite complex and awesome.

Basically, there are two ways of commercially generating steam to drive the blades of a turbine-generator to create electricity. Steam may be generated from heat developed (1) through the combustion of fossil fuels or (2) through nuclear fission. While other methods have evolved that will develop heat (e.g., geothermal energy[1], solar energy, and nuclear fusion[2, 3], this section will be limited to a discussion of these two basic commercial methods.

Figure III-1 presents a flow diagram of a fossil fueled generating plant. The combustible fuel is

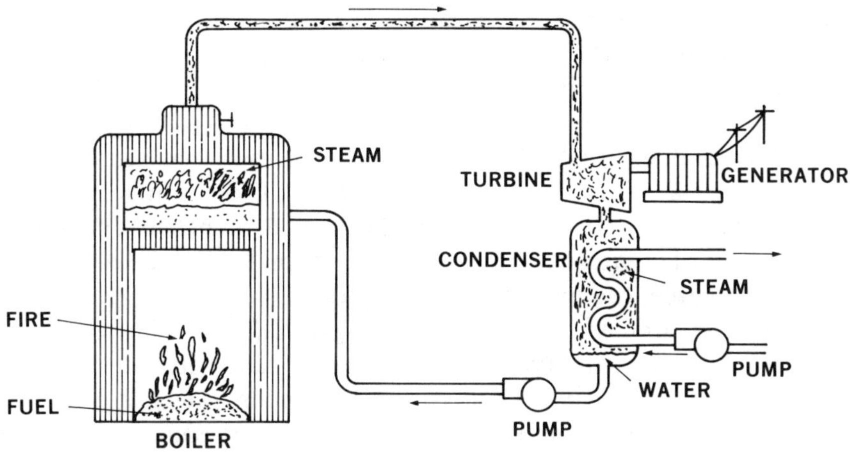

Figure III-1 Diagram of a fossil fuel generating plant.[4]

blown into a furnace to maintain a flame that will heat water that is flowing through tubes in the walls of the boiler. (The name "water-wall boiler" is given to this type of construction.) As explained in Chapter 4, the most efficient means of utilizing this steam is to have it form at high temperatures and pressures.

The steam is conducted to the turbine where it is used to drive the blades of the turbine. The turbine blades are attached to a shaft. Hence, as the steam passes the blades, the blades cause the shaft to rotate. This action transfers energy from the steam to the turbine shaft. As energy is removed from the steam, the pressure and temperature of the steam decreases. (The steam effluent must then be condensed back to water before being processed through the boiler again.) The rotating turbine shaft is used to drive the generator. Electrical energy is produced when a magnetic field is rotated by the turbine shaft past electrical conductors which are present within the generator.

Hence, to generate electricity, energy in the chemical form (fossil fuel) is burned to produce steam, which is converted to mechanical energy by means of a turbine which in turn produces electricity through the rotation of a magnetic field. At each conversion step energy is lost. As a result, the total conversion process from fossil fuel to electricity produces an efficiency that is less than 40%.

An alternate means of generating heat to boil water and create steam is to utilize the decay heat in a nuclear fission reaction.[4-7] This approach is accomplished by using fuel elements and control rods. The fuel elements are generally cylindrical in shape. Interspersed around the fuel elements are control rods that are normally composed of a neutron-absorbing material. These control rods are used to control the flow of heat in the fuel elements by controlling the amount of neutrons available for reaction. The more neutrons that are available, the more heat produced.

The fission process is illustrated in Figure III-2. A neutron causes a nucleus to split into two parts which are called fission products. When formed, they separate at very high speeds. Heat is generated when these products collide with other molecules and are forced to stop their motion. This heat is called "decay heat."

Figure III–3 illustrates a typical boiling water reactor (BWR). Water is used as a medium. The decay heat in the reactor boils the water to

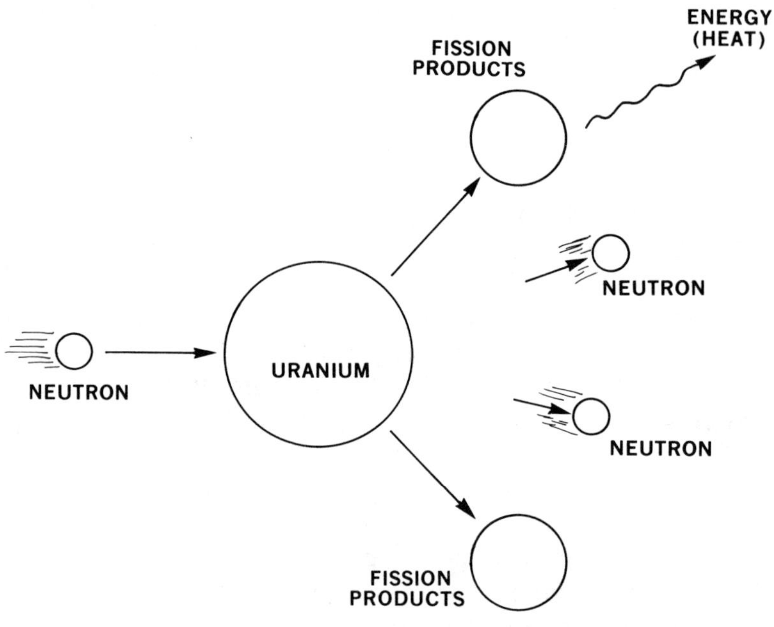

Figure III-2 Nuclear fission process.[6]

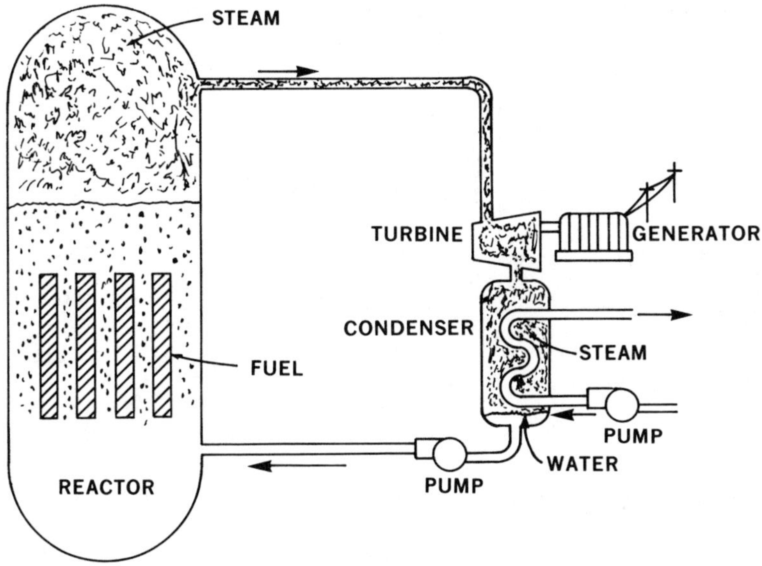

Figure III-3 Diagram of a typical boiling water reactor (BWR).[4]

create steam in a fashion similar to the fossil fueled boiler. The conversions of heat energy in the form of steam to mechanical energy in a turbine and the conversion of mechanical energy to electrical energy in a generator are similar to the conversions which take place in a fossil-fueled plant. The primary areas of caution that have evolved about nuclear fueled plants relate to the radioactivity of the nuclear fuel and the extreme safety precautions which must be taken to ensure the absence of dangerous impacts to the public.

A second type of nuclear fission reactor, in use today, is the pressurized water reactor (PWR) illustrated in Figure III-4. In this system, water is forced through the core under pressures sufficiently high to prevent boiling. The result is more efficient energy conversion. In order to release the heat contained in the water, the water is passed through a steam generator which takes water from a second source and converts it to steam. By separating the water flows in the boiler and the turbine, the likelihood of exposure to radioactivity is reduced.

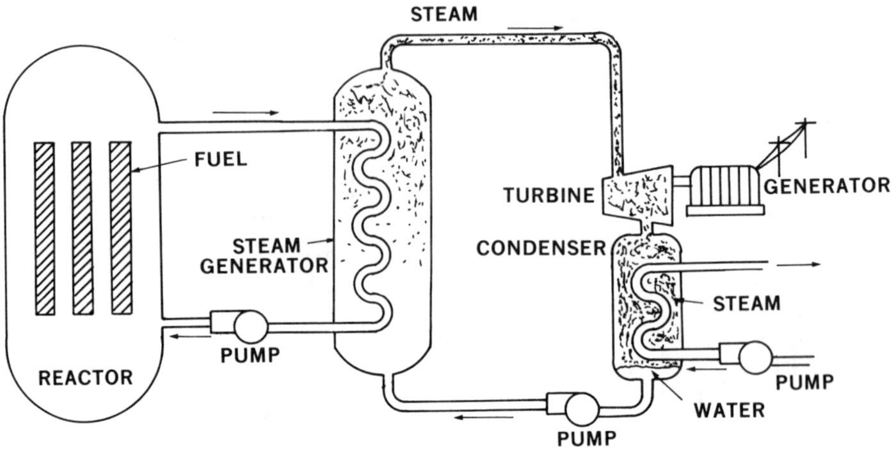

Figure III-4 Diagram of a typical pressurized water reactor (PWR).[4]

Other types of reactors are in use today. For instance, in Canada, a natural-uranium heavy-water reactor (CANDU) is utilized to generate steam.[8] Use of natural uranium bypasses the expensive enrichment step discussed in Chapter 6. High temperature gas reactors (HTGR) are in operation in Colorado and in several European countries. This reactor utilizes a gas mixture as the coolant and heat transfer mechanism and possesses improved efficiencies.

Finally, the "Breeder Reactor" is under development. This reactor uses a special fuel in a process that creates or "breeds" its own fissionable material.[10] This fuel is generally composed of Plutonium 239 and Uranium 238 or Uranium 233 and Thorium 232. Although the theoretical concept has been demonstrated, enormous financing and design problems have prevented commercial demonstration of this type of reactor.

REFERENCES

1. Siever, R., "The Steady State of the Earth's Crust, Atmosphere, and Oceans," *Scientific American,* Volume 230, Number 6, June 1974.
2. Emmett, J. et al, "Fusion Power by Laser Implosion," *Scientific American,* Volume 230, Number 6, June 1974.

3. Steiner, D., "Nuclear Fusion: Focus on Tokamak," *Spectrum,* Volume 14, Number 7, July 1977.

4. "Nuclear Power and the Environment," International Atomic Energy Agency, 1973.

5. Bethe, H., "The Necessity of Fission Power," *Scientific American,* Volume 234, Number 1, January 1976.

6. Hogerton, J., "Nuclear Reactors," U.S. Atomic Energy Commission, 73-605092, 1970.

7. Foster, A. and Write, R., *Basic Nuclear Engineering,* Allyn and Bacon, 1973.

8. McIntyre, H., "Natural Uranium Heavy Water Reactors," *Scientific American,* Volume 233, Number 4, October 1975.

9. Bibbington, W., "Reprocessing of Nuclear Fuels," *Scientific American,* Volume 235, Number 6, December 1976.

10. Mitchell, W. and Turner, S., "Breeder Reactors," U.S. Atomic Energy Commission, 76-611325, 1972.

Appendix IV

THE MODEL STATE UTILITY ENVIRONMENTAL PROTECTION ACT

An Act to provide for the regulation of the location, operation and maintenance of major utility generation and transmission facilities to promote the provision of reliable, abundant and economical energy supply with due regard for the preservation and enhancement of the environment and conservation of scenic, historic, recreational and other natural resources, and for other purposes.

BE IT ENACTED BY THE LEGISLATURE OF THIS STATE:

Section 1. **Short title.** – This Act shall be known, and may be cited, as the "Utility Environmental Protection Act."

Section 2. **Declaration of public policy.** – The Legislature hereby finds and declares that there is at present and will continue to be a growing need for electric and gas public utility services which will require the construction of major new facilities. It is recognized that such facilities cannot be built without in some way affecting the physical environment where such facilities are located. The Legislature further finds that it is essential in the public interest to minimize any adverse effect upon the environment and upon the quality of life of the people of the State which such new facilities might cause. The Legislature further finds that present laws and practices relating to the location of such utility facilities should be strengthened to protect environmental values and to take into account the total cost to society of such facilities. Present laws and practices have in some areas resulted in undue delays in new construction which have increased costs, eventually borne by the people of the State in the form of higher utility rates, and which have threatened the ability of utilities to meet the needs of the people of the State for economic and reliable utility services. Furthermore, the Legislature finds that existing provisions of law may not provide adequate opportunity for individuals, groups interested in conservation and the protection of the environment, State and regional agencies, local governments, and other public bodies to participate in timely fashion in decisions regarding the location and construction of major facilities. The Legislature, therefore, hereby declares that it shall be the purpose of the Act to provide a forum for the expeditious resolution of all matters concerning the location and construction of electric generating plants and electric and gas transmission lines and associated facilities, presently under the jurisdiction of multiple State, regional and local agencies, and all matters of State, regional and local law, in a single proceeding to which access will be open to individuals, groups, State and regional agencies, local governments and other public bodies to enable them to participate in these decisions.

Section 3. **Definitions.** – The following words, when used in this Act, shall have the following meanings, unless otherwise clearly apparent from the context:

(a) The word "Commission" shall mean the Public Service Commission of this state;

(b) The words "major utility facility" shall mean: (1) electric generation and transmission plant designed for, or capable of, operation at a capacity of fifty megawatts or more; (2) an electric transmission line and associated facilities of a design capacity of one hundred twenty-five kilovolts or more; and (3) an intrastate gas transmission line and associated facilities designed to be operated at a hoop stress in excess of 40% of the specified minimum yield strength extending a distance of one mile or more; or is designed to operate at a hoop stress of 20% of specified minimum yield strength or more, and less than 40% of specified minimum yield strength extending a distance of more than ten miles.

(c) The words "commence to construct" shall mean any clearing of land, excavation, or other action that would adversely affect the natural environment of the site or route of a major utility facility, but does not include changes needed for temporary use of sites or routes for non-utility purposes, or uses in securing geological data, including necessary borings to ascertain foundation conditions;

(d) The word "municipality" shall mean any county or municipality within this State;

(e) The word "person" shall include any individual, group, firm, partnership, corporation, cooperative, association, government subdivision, government agency, local government, or other organization; and

(f) The words "public utility" or "utility" shall mean any person engaged in the production, transmission, storage, sale, delivery or furnishing of electricity or gas, or both, for public use.

Section 4. **Certificate of environmental compatability and public need.** – (a) No person shall hereafter commence to construct a major utility facility in the State without first having obtained a certificate of environmental compatibility and public need (hereafter in this Act called a "certificate") issued with respect to such a facility by the Commission. The replacement of an existing facility with a like facility, as determined by the Commission, shall not constitute construction of a major utility facility. Any facility, with respect to which a certificate is required shall be constructed in conformity with such certificate and any terms, conditions and modifications contained therein. A certificate may only be issued pursuant to this Act, provided, however, certificates of public convenience and necessity issued pursuant to other provisions of the public utility laws of this State shall, with respect to facilities to which this Act relates, constitute certificates hereunder if the requirement of this Act have been complied with in proceedings for the issuance thereof.

(b) A certificate may be transferred, subject to the approval of the Commission, to a person who agrees to comply with the terms, conditions and modifications contained therein.

(c) A certificate issued hereunder may be amended as herein provided.

(d) This Act shall not apply to any major utility facility:

(1) For which, prior to the effective date of this Act, an application for the approval of same has been made to any Federal, State, regional or local governmental agency, which possesses the jurisdiction to consider the matters prescribed for finding and determination in Subsection (a) of Section 9 of this Act;

(2) This Act shall not apply to any major utility facility upon which site survey has, prior to the effective date hereof, been commenced, or for which, prior to the effective date hereof, financial commitments have been made by the person proposing to construct same;

(3) For which, prior to the effective date of this Act, a governmental agency has approved the construction of same and such agency has incurred indebtedness to finance all or part of the cost of such construction; or

(4) Over which an agency of the Federal Government has exclusive jurisdiction.

(e) Any person intending to construct a major utility facility excluded from this Act pursuant to Paragraphs (1) or (2) of Subsection (d) of this Section may elect to waive such exclusion by delivering notice of such waiver to the

Commission. This Act shall thereafter apply to each major utility facility identified in such notice from the date of its receipt by the Commission.

Section 5. **Application of certificate.** – (a) An applicant for a certificate shall file with the Commission an application, in such form as the Commission may prescribe, containing the following information: (1) a description of the location and of the major utility facility to be built thereon; (2) a summary of any studies which have been made of the environmental impact of the facility; (3) a statement explaining the need for the facility; (4) a description of any reasonable alternate location or locations for the proposed generating facility, a description of the comparative merits and detriments of each location submitted, and a statement of the reasons why the primary proposed location is best suited for the facility; and (5) such other information as the applicant may consider relevant or as the Commission may by regulation or order require. A copy or copies of the studies referred to in clause (2) above shall be filed with the Commission, if ordered and shall be available for public inspection.

(b) Each application shall be accompanied by sufficient copies for service by the Commission on the chief executive officer of each municipality and the head of each government agency, charged with the duty of protecting the environment or of planning land use, in the area in which any portion of such facility is to be located, both as primarily and as alternatively proposed. The copy of such application shall be accompanied by a notice specifying the date on or about which the application is to be filed.

(c) Each application shall also be accompanied by a summary thereof which the Commission shall cause to be published at applicant's expense, as public notice to persons residing in the local governments entitled to receive notice under Subsection (b) of this Section. Such summary shall be published promptly after filing of the application in a newspaper published or having circulation in such local government area or areas.

(d) Inadvertent failure of service on, or notice to, any of the municipalities, government agencies or persons identified in Subsections (b) and (c) of this Section may be cured pursuant to orders of the Commission designed to afford them adequate notice to enable their effective participation in the proceeding. In addition, the Commission may, after filing, require the applicant to serve notice of the application, or copies thereof or both upon other persons, and file proof thereof, as the Commission may deem appropriate.

(e) An application for an amendment of a certificate shall be in such form and contain such information as the Commission shall prescribe. Notice of such an application shall be given as set forth in Subsections (b) and (c) of this Section.

Section 6. **Hearing on application for certificate; amendment.** – (a) Upon the receipt of an application complying with Section 5, the Commission shall promptly fix a date for the commencement of a public hearing thereon, not less than thirty nor more than sixty days after such receipt, and shall conclude the proceedings as expeditiously as practicable. The testimony presented at such hearing may be presented in writing or orally, provided that the

Commission may make rules designed to exclude repetitive, redundant or irrelevant testimony.

(b) On an application for an amendment of a certificate, the Commission shall hold a hearing in the same manner as a hearing is held on an application for a certificate if the proposed change in the facility would result in any material increase in any environmental impact of the facility or a substantial change in the location of all or a portion of such facility other than as provided in the alternates set forth in the application.

Section 7. **Parties to certification proceedings.** – (a) The parties to a certification proceeding shall include:

(1) The applicant;

(2) Each municipality and government agency entitled to receive service of a copy of the application under Subsection (b) of Section 5 of this Act, if it has filed with the Commission a notice of intervention as a party, within fifteen days after the date it was served with a copy of the application; and

(3) Any person residing in a municipality entitled to receive service of a copy of the application under Subsection (b) of Section 5 of this Act; any domestic non-profit organization, formed in whole or in part to promote conservation or natural beauty, to protect the environment, personal health or other biological values, to preserve historical sites, to promote consumer interests, to represent commercial and industrial groups, or to promote the orderly development of the areas in which the facility is to be located; or any other person, if such a person or organization has petitioned the Commission for leave to intervene as a party, within fifteen days after the date given in the published notice as the date for filing the application, and if such petition has been granted by the Commission for good cause shown.

(b) Any person may make a limited appearance in the proceeding by filing a verified statement of position within thirty days after the date given in the published notice as the date for filing the application. No person making a limited appearance shall be a party or shall have the right to present oral testimony or argument or cross-examine witnesses.

(c) The Commission may, in extraordinary circumstances for good cause shown, grant a petition for leave to intervene as a party to participate in subsequent phases of the proceeding, filed by a municipality, government agency, person or organization who is identified in Paragraphs (2) or (3) of Subsection (a) of this Section, but who failed to file a timely notice of intervention or petition for leave to intervene, as the case may be.

Section 8. **Conduct of the hearing.** – A record shall be made of the hearing and all testimony taken and the cross-examination thereon. Rules of evidence, as specified by the Commission, shall apply to the proceeding. The Commission may provide for the consolidation of the representation of parties having similar interests.

Section 9. **The decision.** – (a) The Commission shall render a decision upon the record either granting or denying the application as filed, or granting it upon such terms, conditions, or modifications of the construction, operation or maintenance of the major utility facility as the Commission may deem appropriate. The Commission may not grant a certificate for the construction, operation and maintenance of a major utility facility, either as proposed or as modified by the Commission, unless it shall find and determine:

(1) The basis of the need for the facility;

(2) The nature of the probable environmental impact;

(3) That the facility represents the minimum adverse environmental impact, considering the state of available technology and the nature of economics of the various alternatives, and other pertinent considerations;

(4) In the case of an electric transmission line, what part, if any, of the line shall be located underground; that such facility is consistent with regional plans for expansion of the electrical power grid systems serving the State and interconnected utility systems; and that such facilities will serve the interests of electric system economy and reliability;

(5) That the location of the facility as proposed conforms to applicable State, regional and local laws and regulations issued thereunder, except that the Commission may refuse to apply any regional or local law or regulation if it finds that, as applied to the proposed facility, such law or regulation is unreasonably restrictive in view of the existing technology, or factors of cost or economics, or of the needs of consumers whether located inside or outside of the directly affected government subdivisions; and

(6) That the facility will serve the public interest, convenience and necessity.

(b) If the Commission determines that the location of all or a part of the proposed facility should be modified, it may condition its certificate upon such modification, provided that the municipalities, and persons residing therein, affected by the modification, shall have been given reasonable notice thereof.

(c) A copy of the decision and any opinion issued therewith shall be served upon each party.

Section 10. **Opinion to be issued with decision.** – In rendering a decision on an application for a certificate, the Commission shall issue an opinion stating its reasons for the action taken. If the Commission has found that any regional or local law or regulation, which would be otherwise applicable, is unreasonably restrictive pursuant to Paragraph (5) of Subsection (a) of Section 9 of this Act, it shall state in its opinion the reasons therefor.

Section 11. **Rehearing: judicial review.** – (a) Any party aggrieved by any decision issued on an application for a certificate may apply for a rehearing within fifteen days after issuance of the decision. Any party aggrieved by the final decision of the Commission on rehearing may obtain judicial review thereof by filing of a petition in a State court of competent jurisdiction within fifteen days after the issuance of such final decision. Upon receipt of such petition, the Commission shall forthwith deliver to the court a copy of the written transcript of the record of the proceeding before it and a copy of its decision and opinion entered therein which shall constitute the record on judicial review. A copy of such transcript, decision and opinion shall remain on file with the Commission and shall be available for public inspection.

(b) The court in the proceeding shall have the power to grant such relief as it deems just and proper, and to make and enter an order enforcing, modifying and enforcing as so modified, remanding for further specific evidence or findings, or setting aside in whole or in part, such decision of the Commission. The appeal shall be heard on the record without requirement of reproduction. No objection that has not been urged by the aggrieved party in his application for rehearing before the Commission shall be considered by the court, unless the failure or neglect to urge such objection shall be excused because of extraordinary circumstances. The findings of fact on which such decision is based shall be conclusive if supported by substantial evidence on the record considered as a whole. The jurisdiction of such court shall be exclusive and its judgment and order shall be final, subject only to review by the highest appellate court of the state in the same manner as provided for appeals to such court in other proceedings. All such proceedings shall be heard and determined by the courts as expeditiously as practicable and with lawful precedence over other matters.

(c) The grounds for and the scope of review of the court shall be limited to whether the decision and opinion of the Commission is:

(1) In conformity with the constitution and the laws of the State and of the United States;

(2) Supported by substantial evidence in the record;

(3) Made in accordance with the procedures set forth in this Act or established by order, rule or regulation of the Commission; and

(4) Arbitrary, capricious or an abuse of discretion.

Section 12. **Jurisdiction of courts.** – Except as expressly set forth in Section 11 of this Act, no court of this State shall have jurisdiction to hear or determine any issue, case or controversy concerning any matter which was or could have been determined in a proceeding before the Commission under this Act or to stop or delay the construction, operation or maintenance of a major utility facility except to enforce compliance with this Act or the provisions of a certificate issued hereunder.

Section 13. **Powers of local governments and State agencies.** – Notwithstanding any other provision of law, no State or regional agency, or municipality or other local government may require any approval, consent, permit, certificate or other condition for the construction, operation or maintenance of a major utility facility authorized by a certificate issued pursuant to the provisions of this Act; except that nothing herein shall prevent the application of State laws for the protection of employees engaged in the construction, operation or maintenance of such facility.

Section 14. **Joint hearings and orders.** – The Commission, in the discharge of its duties under this Act or any other Act, is authorized to make joint investigations, hold joint hearings within or without the State, and issue joint or concurrent orders in conjunction or concurrence with any official or agency of any

State or of the United States, whether in the holding of such investigations or hearings, or in the making of such orders, the Commission shall function under agreements or compacts between States or under the concurrent power of states to regulate interstate commerce, or as an agency of the United States, or otherwise. The Commission, in the discharge of its duties under this Act, is further authorized to negotiate and enter into agreements or compacts with agencies of other States, pursuant to any consent of Congress, for cooperative efforts in certifying the construction, operation and maintenance of major utility facilities in accord with the purposes of this Act and for the enforcement of the respective State laws regarding same.

Section 15. **Effective date.** – This Act shall become effective . . . . .

Bibliography

BOOKS

1. Foster, A. and Wright, R., *Basic Nuclear Engineering,* Allyn and Bacon, 1973.

2. Healy, T., *Energy, Electric Power, and Man,* Boyd and Fraser, 1974.

3. Karam, R. and Morgan, K., *Environmental Impact of Nuclear Power Plants,* Pergamon Press, 1976.

4. Sagan, L. (Ed.), *Human and Ecological Effects of Nuclear Power Plants,* Charles C. Thomas, 1974.
   (a) Jensen, L., "The Effects of Thermal Discharge into Surface Waters," pp. 440–479.

5. Schurr, S., *Energy, Economic Growth and the Environment,* Johns Hopkins University Press, 1971.

6. Solberg, C., *Oil Power,* Mason/Carter, 1976.

CONFERENCE PROCEEDINGS

1. *ASCE Annual Conference of 1972,* Calvert, J. and Heilman, W., "A New Approach to PPS."
2. *American Power Conference,* May 1973, Nassikas, J., "National Energy Policy: Directions and Developments."
3. *Atomic Industrial Forum Conference on U.S. Energy Policy,* January 1977, "Impacts of NEPA and Other Court Actions on the Nuclear Option."
4. *Conference on Constructing Social Impact Statements for Power Plant Siting,* September 1975, Ohio State University.
5. *Conference on Nuclear Power Plant Siting,* August 26–28, 1974, American Nuclear Society.
   (a) Elkins, M. and DiNunno, J., "A Regional Siting for Thermal Power Plants in the State of Ohio."
   (b) Fisher, J. and Ahmed, R., "A Systematic Approach to Evaluate Sites for Nuclear Power Plants."
   (c) Norris, J., "Selected Environmental Considerations and their Measuring Parameters."
   (d) West, L. and Dyar, K., "The Methodology and Criteria Used for a Nuclear Power Plant Siting Study."
6. *EIS Conference,* August 1973, McCormick, J., "Content and Mechanics of Environmental Impact Statements."
7. *International Energy Engineering Congress,* November 1975, Dul, E. et al, "Principles of Quantitative Environmental Impact Analysis for Proposed Power Plants."
8. *Second Iranian Congress of Civil Engineers,* May 3–7, 1976, Pahlavi University, Hagerty, J. and Conner, D., "Environmental Considerations in Power Plant Siting."
9. *Second State-Federal Power Plant Siting Conference,* June 16–18, 1976, Sponsored Jointly by the U.S. Nuclear Regulatory Commission and the National Governor's Conference Energy Conference.
10. *Seminar on the Engineering Aspects of Siting and Operating Power Plants,* 1973, Sponsored by the U.S. Department of Interior and the Atomic Industrial Forum, Inc.
11. *1975 Federal-State Conference on Power Plant Siting,* July 1975, U.S. Nuclear Regulatory Commission, Office of International and State Programs.

EDUCATIONAL MATERIALS

1. "Energy and the Law: Problems and Challenges of the Late 70's," American Law Institute – American Bar Association Study Course, 1977.

GOVERNMENT DOCUMENTS

Atomic Energy Commission, U.S.

1. Hogerton, J., "Nuclear Reactors," Division of Technical Information, 1963.
2. Hogerton, J., "Nuclear Reactors," 73–605092, 1970.
3. Mitchell, W. and Turner, S., "Breeder Reactors," Office of Information Services, 76–611325, 1972.
4. Ramsey, W. and Reed, P., "Land Use and Nuclear Power Plants Case Studies of Siting Problems," Directorate of Regulatory Standards, WASH 1319, October 1974.
5. "Toxicity of Power Plant Chemicals to Aquatic Life," WASH 1249, June 1973.

Canada

1. Porter, A. (Chairman), "Royal Commission on Electric Power Planning," First Report, Royal Commission on Electric Power Planning, May 1976.
2. "Story of Thermal Discharge Diffusion for the Canadian Great Lakes Shore-Line," Department of the Environment Center for Inland Waters, Prepared by Acres Consulting Services, Ltd., March 1974.

Energy Research and Development Administration, U.S.

1. "A National Plan for Energy Research and Development and Demonstration: Creating Energy Choices for the Future," April 1976.
2. "Creating Energy Choices for the Future," 1975.
3. "Creating Energy Choices for the Future," 1976.
4. Dukert, J., "High-level Radioactive Waste: Safe Storage and Ultimate Disposal," 1975.
5. Olsen, M. and Merwin, D., "Toward a Methodology for Constructing Social Impact Statements Using Quality of Social Life Indicators," Prepared by Battelle Pacific Northwest Laboratory, 1976.
6. Simmons, H., "The Economics of America's Energy Future," 1975.

Environmental Protection Agency, U.S.

1. "Analysis and Control of Thermal Pollution," PB 237 617, August 1974.
2. "Development Document for Best Technology Available for the Location, Design, Construction, and Capacity of Cooling Water Intake Structures for Minimizing Adverse Environmental Impact," April 1976.
3. "Krypton 85, A Review of the Literature and An Analysis of Radiation Hazards," Office of Research and Monitoring, January 1972.

4. Nelson, G., "Water Recycle/Reuse Possibilities: Power Plant Boiler and Cooling Systems," December 1974.

5. Nelson, G., "Technical and Economic Evaluation of Cooling Systems Blowdown Control Techniques," Office of Research and Development, November 1973.

6. "Observations of Airborne Tritium Waste Discharge from a Nuclear Fuel Reprocessing Plant," Office of Radiation Programs – Field Operations Division, February 1973.

Federal Energy Administration, U.S.

1. "Monthly Energy Review," July, August, and September 1976.
2. "Structural Reform in the Electric Power Industry," G–761323, 1976.

Federal Power Commission, U.S.

1. "Factors Affecting the Electric Power Supply 1980–85," Bureau of Power, 1976.

Florida, State of
Tallahassee, Florida

1. "Ten-year Site Plans for Electrical Generating Facilities and Associated Transmission Lines," Department of Administration Division of State Planning, September 24, 1973.

Government Printing Office, U.S.

1. "1976 National Energy Outlook," Document No. 0–204–986, 1976.
2. Doub, W. (Chairman), "Federal Energy Regulation – An Organizational Study," Federal Energy Regulation Study Team, Document No. 5210-00386, April 1974.
3. *Energy Alternatives: A Comparative Analysis,* Stock No. 041-011-00025-4, Catalog No. PRES 14.2:EN2, May 1975.
4. "MERES and the Evaluation of Energy Alternatives," Stock No. 041-011-00026-2, Catalog No. PREX 14.2:M42.

Health, Education, and Welfare, (U.S.) Department of

1. "Air Quality Criteria for Nitrogen Oxides," Public Health Service, National Air Pollution Control Administration, January 1969.
2. "Air Quality Criteria for Particulate Matter," Public Health Service, National Air Pollution Control Administration, January 1969.

3. "Air Quality Criteria for Sulfur Oxides," Public Health Service, National Air Pollution Control Administration, January 1969.

4. "An Investigation of Airborne Radioactive Effluent from an Operating Nuclear Fuel Reprocessing Plant," Public Health Service, July, 1970.

5. Hindaul, I., "Air Pollution Injury to Vegetation," Public Health Service, National Air Pollution Control Administration, 1970.

Interior, (U.S.) Department of

1. "United States Energy through the Year 2000 (Revised)," Bureau of Mines, December 1975.

Kentucky, Commonwealth of
Frankfort, Kentucky

1. "Energy for Industry in Kentucky," Department of Commerce, Research and Planning Division, 1970.

Maryland, State of
Annapolis, Maryland

1. "Addendum to PPSE 1-2, Power Plant Site Evaluation, Brandon Shores Site," Department of Natural Resources Power Plant Siting Program, Prepared by Johns Hopkins University, January 18, 1973.

2. "Calculated Program Power Plant Impact on Baltimore Air Shed," Department of Natural Resources Power Plant Siting Program, Prepared by Research Institute for Advanced Studies, January 1973.

3. "Douglas Point, A Story of Circulation and Mixing," Department of Natural Resources Power Plant Siting Program, Prepared by Chesapeake Bay Institute, May 1975.

4. "Power Plant Cumulative Environmental Impact Report," Department of Natural Resources Power Plant Siting Program, PPSP-CEIR-1, September 1975.

5. "Power Plant Site Evaluation, Brandon Shores Site," Department of Natural Resources Power Plant Siting Program, Prepared by Johns Hopkins University, PPSE-12, November 1972.

6. "Power Plant Site Evaluation Interim Report – Douglas Point Site," Vol. 2, Department of Natural Resources Power Plant Siting Program, Prepared by Johns Hopkins University, May 1974.

7. "Power Plant Site Evaluation Report, Dickerson Site," Department of Natural Resources Power Plant Siting Program, Prepared by Johns Hopkins University, Addendum to PPSE 3-1, January 1974.

8. "Power Plant Site Evaluation Report, Douglas Point Site," Vol. 1 and 2, Department of Natural Resources Power Plant Siting Program, Prepared by Johns Hopkins University, May 1974.

9. "Preliminary Investigations of a Potential Power Plant Site in Cecil County, Md.," Vol. 1 – Summary, Department of Natural Resources Power Plant Siting Program, Prepared by Engineering-Science, Inc., February 1974.

Massachusetts, Commonwealth of
Boston, Massachusetts

1. "The Need for Generation in New England: A Strategy to Develop a Modeling Approach for Evaluating Proposed Electrical Facilities," Energy Facilities Siting Council, April 1977.

National Science Foundation, U.S.

1. "Geothermal Energy Program," March 1976.

Nuclear Regulatory Commission, U.S.

1. "Alternative Financing Methods," Office of State Programs, NUREG-0204, March 1977.
2. "Detailed Study Plan Efficiency in Federal/State Siting Actions," Office of State Programs, NUREG-0128, October 1976.
3. Dukert, J., "High-level Radioactive Waste: Safe Storage and Ultimate Disposal," 1975.
4. "Early Site Reviews for Nuclear Power Facilities," Office of Nuclear Reactor Regulation, NUREG-0180, May 1977.
5. "Environmental Effects of Using Coal for Generating Electricity," Prepared by Argonne National Laboratory, NUREG-0252, June 1977.
6. "Environmental Planning and the Siting of Nuclear Facilities: The Integration of Water, Air, Coastal, and Comprehensive Planning into the Nuclear Siting Process," Office of State Programs, NUREG-0199, February 1977.
7. Fox, C., "Radioactive Wastes," 1969.
8. Hogerton, J., "Nuclear Reactors," 1970.
9. "Improving Regulatory Effectiveness in Federal/State Siting Actions," Office of State Programs, NUREG-0195, 1977.
10. "Nuclear Energy Center Site Survey – 1975, Part III of V, Technical Considerations, NUREG-0001, January 1976.
11. "Nuclear Power Plant Licensing: A New England Perspective," Office of State Programs, NUREG-0202, March 1977.
12. "Radioactive Materials Released from Nuclear Power Plants," May 1976.

13. Rasmussen, H., "Reactor Safety Study: Assessment of Accidental Risks in U.S. Commercial Nuclear Power Plants," WASH 1400, 1976.

14. "Regulatory Guide 4.2, Revision 1: Preparation of Environmental Reports for Nuclear Power Station," January 1975.

15. "Report to Congress on Abnormal Occurrences October-December 1975," March 1976.

16. "Socioeconomic Impacts: Nuclear Power Station Siting," NUREG–0150, June 1977.

17. "Standard Format of Safety Analysis Reports for Nuclear Power Plants," February 1974.

18. "State and Local Planning Procedures Dealing with Social and Economic Impacts from Nuclear Power Plants," Office of State Programs, NUREG–0203, January 1977.

19. "State Regulatory Activity Involved in Need for Power," Office of State Programs, NUREG–0197, April 1977.

20. "Success Factor Evaluation Panel," Office of State Programs, NUREG–0196, March 1977.

Ohio, State of
Columbus, Ohio

1. Felman, D., et al, "CAPS: Comparative Analysis of Power Siting in the U.S.," Power Siting Commission, June 1977.

2. "Guidelines for Electric Utilities in Assessing Environmental Effects," Department of Health, Prepared by Environmental Research Associates, 1970.

3. "Preliminary Criteria for Evaluating Major Utility Facility Applications," Power Plant Siting Commission, Prepared by URS Corp., 1975.

4. Skipton, E. and Wayland, R., "Comments on 'Economics of Nuclear Power'," Public Utilities Commission Finance and Economics Section, December 5, 1975.

5. "State of Ohio Environmental Assessment Program Information Manual," Department of Natural Resources, 1972.

6. Traub, R., "An Investigation on the Socio-Economic Impact of the Nuclear Power Plants on Small Communities," Power Plant Siting Commission, 1974.

Other

1. "National Environmental Policy Act of 1969," Public Law 91–190, 91st Congress, S.1075, January 1, 1970, Title I, Section 102, Part C.

Pennsylvania, State of
Harrisburg, Pennsylvania

1. "Energy Facility Siting: A Comprehensive Program for Pennsylvania," Joint State Government Commission, April 1977.
2. "Power Plant Siting Policy Alternatives for Pennsylvania," Department of Engineering and Public Affairs, 1975.

Texas, State of
Austin, Texas

1. "Electric Power in Texas," Office of the Governor – Division of Planning Coordination – Advisory Committee on Plant Siting, 1972.

Wisconsin, State of
Madison, Wisconsin

1. "Local Impacts of Power Plant Siting in Wisconsin," Public Service Commission, December 1975.

PERIODICALS

Business Week

1. "Energy Conservation's Impact on R & D," June 27, 1977.
2. "The Government Gambles on the Centrifuge," May 30, 1977.
3. "U.S. Nuclear Power May Be on the Wrong Track," April 18, 1977.
4. "Will Energy Conservation Throttle Economic Growth?," April 25, 1977.

Chemical and Engineering News

1. "Fusion Focus Shifts to Practical Aspects," Vol. 54, No. 42, October 11, 1976.
2. "Geothermal Energy Promising but Risky," Vol. 54, No. 42, October 11, 1976.

Electrical World

1. "California Utilities View Coastal Zoning Differently," November 1, 1976.
2. "The Electric Century 1874–1974," June 1, 1974.

Environment

1. Novick, S., "The Electric Power Industry," Vol. 17, No. 8, November 1975.

Forbes

1. Cook, J., "The Invisible Crisis," July 15, 1976.

Foreign Affairs

1. Wilson, C., "A Plan for Energy Independence," Vol. 51, No. 4, July 1973.

Fortune

1. Alexander, T., "Industry Can Save Energy Without Stunting Its Growth," May 1977.

Nuclear News

1. Wilson, R., "Some Perspectives on the Energy Question," February 1976.

Nuclear Technology

1. Chester, C., and Chester, R., "Civil Defense Implications of the U.S. Nuclear Power Industry During a Large Nuclear War in the Year 2000," Vol. 31, December 1976.
2. Evans, V., "Recent State Laws Regulating Power Facility Siting," Vol. 31, December 1976.

Forbes

1. Cole, S., "Ohio River Basin Energy Studies," October/November 1976.

Scientific American

1. Bethe, H., "The Necessity of Fission Power," Vol. 230, No. 1, January 1976.
2. Bibbington, W., "Reprocessing of Nuclear Fuels," Vol. 235, No. 6, December 1976.
3. Emmett, J. et al, "Fusion Power by Laser Implosion," Vol. 230, No. 6, June 1974.
4. McIntyre, H., "Natural Uranium Heavy Water Reactors," Vol. 233, No. 4, October 1975.
5. Siever, R., "The Steady State of the Earth's Crust, Atmosphere, and Oceans," Vol. 230, No. 6, June 1974.

Spectrum, IEEE

1. "Energy Report," Vol. 14, No. 7, July 1977.
2. Steiner, D., "Nuclear Fusion: Focus on Tokamak," Vol. 14, No. 7, July 1977.

Washington Law Review

1. Lippek, H., "Power and the Environment: A Statutory Approach," Vol. 47, pp. 35–72, 1971.

2. Rogers, W., "Siting Power Plants in Washington State," Vol. 47, pp. 9–33, 1971.

REPORTS (NONGOVERNMENTAL)

1. "6th Annual Review of Overall Reliability and Adequacy of the North American Bulk Power System," National Electric Reliability Council, 1976.

2. "A Summary of Program Emphasis for 1976," Electric Power Research Institute.

3. "A Survey of Public Leadership Attitudes toward Nuclear Power Development in the United States," EBASCO Services, Inc., August 1975.

4. Ball, R. et al, "California's Electricity Quandary: 11. Planning for Power Plant Siting," RAND, September 1972.

5. Beck, R., "Dialogue on Power Demand," The Urban Studies Center, Portland State University, December 1972.

6. Beck, R., et al, "Underground Nuclear Power Plant Siting," California Institute of Technology, 1972.

7. Chang S. and Snead, R., "An Annotated Bibliography: Social and Economic Factors Associated with Electric Power Generating Stations," Atomic Industrial Forum, October 1976.

8. Coe, R. (Chairman), "Environmental Guidelines," Western System Coordinating Council Environmental Committee, December 3, 1971.

9. "Evaluation of Environmental Effects of a Natural Draft Cooling Tower at the Davis-Bosse Nuclear Power Station," NUS Corporation, July 1971.

10. "Fossil and Nuclear Fuel for the Electric Utility, Generation Requirements and Constraints 1976–1985," National Electric Reliability Council, 1976.

11. Geitner, N. and Broome, K., "Systematic Methods for Power Plant Siting Studies," Presented at the American Nuclear Society Power Division Topical Meeting, August 26–28, 1974, Published by Gilbert Commonwealth Associates.

12. "General Environmental Guidelines for Evaluating and Reporting the Effects of Nuclear Power Plant Site Preparation, Plant and Transmission Facility Construction," Prepared by Hittman Associates, Inc., for the Atomic Industrial Forum, February 1974.

13. Hauck, F., "Unneeded Generating Plants in Ohio River Valley," Unpublished Report Prepared by Author in Louisville, Kentucky, November 1976.

14. Jones, N., "Report on the Socio-Economic Impact Issues Working Group," State/Federal Siting Commission Committee Meeting of June 16, 1976.

15. Kendall, H., "Nuclear Power Risks – A Review of the Report of the American Physical Society's Study Group on Light Water Reactor Safety," Union of Concerned Scientists, Cambridge, Massachusetts, June 1975.

16. Kendall, H. and Mowgler, S., "Preliminary Review of the AEC Reactor Safety Study," Joint Review Committee of the Sierra Club (San Francisco, California) and the Union of Concerned Scientists (Cambridge, Massachusetts), November 1974.

17. Krishna, P. and Douglas, R., "Nuclear Power Plant Siting Considerations and Experiences," Public Service Electric and Gas Company, 1974.

18. "Laws and Procedures of Power Plant Siting in New England and New York (Revised)," New England River Basin Commission, March 1972.

19. Lees, L., "People, Power, Pollution," California Institute of Technology, September 1, 1971.

20. Miller, D., "How to Find a Plant Site Without Losing Your Mind," American Electric Power System, 1969.

21. "National Survey on Electric Utilities Security," Burns Security Institute, 1976.

22. "Nuclear Power and the Environment," International Atomic Energy Agency, Vienna, 1973.

23. Perry, R. and Rutledge, J., "Siting Considerations for Offshore Nuclear Power Plants," Dames and Moore, September 1973.

24. Petruschell, R. and Salter, R., "Electricity Generating Cost Model for Comparison of California Power Plant Siting Alternatives," RAND, January 1973.

25. Philbin, T., "Imputs of NEPA and Other Court Actions on Fossil Fuel Option," Ecological Analysts, Inc., January 11, 1977.

26. "Potential Fuel Shifts," Stone and Webster Management Consultants, Inc., 1975.

27. "Power Plant Siting in the United States," Southern Interstate Nuclear Board, September 1974.

28. "Power Plant Siting in the United States," Southern Interstate Nuclear Board, June 1976.

29. "Reactions of the Public and Thought Leaders to a Floating Nuclear Electric Generating Plant," Westinghouse Electric Corp., Prepared by Opinion Research Corp., January 1972.

30. "Resource and Land Investigations (RALI) Program: Methodologies for Environmental Analysis, Vol. 2, Power Plant Siting," MITRE Corporation, MTR 6740, August 1974.

31. "Review of Power Planning in the Pacific Northwest," Power Planning Committee, Pacific Northwest River Basins Commission, Calendar Year 1971, January 1972.

32. "Siting of Nuclear Facilities," International Atomic Energy Agency, Vienna, 1975.

33. "The Importance of Closing the Nuclear Fuel Cycle," Atomic Industrial Forum, October 1, 1976.

34. "The Need for Energy Facility Sites in the United States: 1975-1985 and 1985-2000," Environmental Policy Institute, June 1975.

35. Weinberg, A., "Energy Policy and Energy Projections: The Case of a Nuclear Moratorium," Institute for Energy Analysis, Oak Ridge Associated Universities, Oak Ridge, Tennessee, January 11, 1977.

SPEECHES

1. Gambs, G., "The Energy Crisis in the United States," Presented to the NAM Joint Policy Committee Conference, March 1973.

Index

Index